MENTAL ILLNESS STIGMA AND THE MORAL AND SOCIAL COMMUNITY

Although mental health is a better-understood, more widely discussed topic in our society today, a degree of stigmatization persists, especially in severe cases, with links to homelessness, job loss, poverty, and human rights. It is still present in environments such as the workforce, health-care settings, and educational environments, and is often internalized by the sufferers themselves. This book provides a philosophical account of what mental illness stigma is, why it persists, what harms it causes to people subject to public stigma or who internalize stigma in themselves, and what can be done about it. It analyzes the process of stigmatization, both public and internalized, in the twenty-first-century Western culture, especially in the United States – including the process of stereotyping, the expressive harm of stereotypes, the role of social norms in creating adaptive preferences and shaping behavior, the moral distancing and status loss involved with social exclusion and dehumanization, and the harm of discrimination.

ABIGAIL GOSSELIN is Professor of Philosophy at Regis University. She is the author of *Global Poverty and Individual Responsibility* (2009), *Humanizing Mental Illness* (2021), *Mental Patient* (2022), *Mental Health Resilience* (2024), and *Self-Care in the Face of Stigma* (forthcoming).

MENTAL ILLNESS STIGMA AND THE MORAL AND SOCIAL COMMUNITY

ABIGAIL GOSSELIN
Regis University

CAMBRIDGE
UNIVERSITY PRESS

CAMBRIDGE UNIVERSITY PRESS

Shaftesbury Road, Cambridge CB2 8EA, United Kingdom

One Liberty Plaza, 20th Floor, New York, NY 10006, USA

477 Williamstown Road, Port Melbourne, VIC 3207, Australia

314–321, 3rd Floor, Plot 3, Splendor Forum, Jasola District Centre, New Delhi – 110025, India

Cambridge University Press is part of Cambridge University Press & Assessment, a department of the University of Cambridge.

We share the University's mission to contribute to society through the pursuit of education, learning and research at the highest international levels of excellence.

www.cambridge.org
Information on this title: www.cambridge.org/9781009680448

DOI: 10.1017/9781009680431

When citing this work, please include a reference to the DOI 10.1017/9781009680431

First published 2026

Cover image: gremlin / E+ via Getty Images.

A catalogue record for this publication is available from the British Library

Library of Congress Cataloging-in-Publication Data
NAMES: Gosselin, Abigail, 1977– author
TITLE: Mental illness stigma and the moral and social community / Abigail Gosselin, Regis University, Colorado.
DESCRIPTION: Cambridge, United Kingdom : Cambridge University Press, 2026. | Includes bibliographical references and index.
IDENTIFIERS: LCCN 2025034027 | ISBN 9781009680448 hardback | ISBN 9781009680455 paperback | ISBN 9781009680431 ebook
SUBJECTS: LCSH: Mental illness – Public opinion. | Mental illness – Social aspects | Stigma (Social psychology) | Stereotypes (Social psychology) | Discrimination against the mentally ill.
CLASSIFICATION: LCC RC455.2.P85 G67 2026
LC record available at https://lccn.loc.gov/2025034027

ISBN 978-1-009-68044-8 Hardback

To Derrick

Contents

Acknowledgments

Writing this book has been a labor of love and commitment. A couple of years ago, I was working on a different manuscript when I realized that mental illness stigma was a more pressing topic in that moment, and I felt compelled to change my research focus. I was especially moved by observing the stigma of people with severe mental illness, especially those who are homeless, and the internalized stigma that many people project onto themselves, especially young adults who pay a lot of attention to social media and are self-conscious and worried about others' judgments of them. I thought it was important to explore the various harms these stigmas cause so that we could have a better understanding of the way stigma operates, which can help us better understand how to address it.

This book would not be possible without the support of my family and my colleagues at Regis University. I thank my husband, Derrick Belanger; my daughters, Rhea and Phoebe Belanger; my parents, Marc and Deb Gosselin; my sister and her partner, Joanne Gosselin and Jason Carter; my mother-in-law, Ellen Belanger; and my brother-in-law and sister-in-law, Brian and Traci Belanger for their constant love and support. I also thank my colleagues in the Philosophy Department at Regis University for their encouragement and support, especially Karen Adkins, whose championing of my work has been pivotal in maintaining a positive attitude in a climate of uncertainty and change at my school. I am lucky to work in an environment that supports the many facets of my work.

Introduction

When I first started writing about mental illness fifteen years ago, mental health issues were something one didn't normally talk about. Mental health was not a subject on many people's radars, and experience with mental illness symptoms and suicidality were usually something one tried to hide from others. At that time, the moral imperative seemed to me to be to raise mental health awareness. As a sufferer of many mental illnesses over the years, including major depression, generalized anxiety disorder, eating disorders, bipolar disorder (with depressive, manic, and mixed episodes), and (more recently) schizoaffective disorder, I thought it was my duty to try to raise awareness of mental health issues to get people to talk about them publicly. As a professor, I shared my own experiences with my students when relevant in classes; I organized some mental health awareness events on my campus; and I wrote about my experiences in various venues. It seemed that talking about mental health problems was key to addressing them.

Fifteen years later, the landscape has changed. Mental health is on everyone's radar. People commonly talk about mental illness and mental health and well-being. Campuses regularly host mental illness awareness events, and students commonly share about their own experiences with their professors. Mental health crisis and suicide hotlines have proliferated, and more and more resources are directed toward making therapy and medication more available to more people. People are talking about mental health and mental illness. Nonetheless, rates of mental illness and despair have increased,[1] and mental illness stigma endures. Talking about the problem did not solve it.

[1] The rate of anxiety in the United States increased from 15.6 percent of adults in 2019 to 18.2 percent in 2022. During this same time frame, the rate of depression increased from 18.5 percent of adults to 21.4 percent. Rates of anxiety and depression were highest among those aged eighteen to twenty-nine, more elevated in women than in men, greater in those with only a high school education, and higher in those with low socioeconomic status (US Centers for Disease Control and Prevention 2024).

While people are more aware of mental illness stigma than ever before, stigma still proliferates. One might think that the fact that mental health problems are so common, that people are talking about mental health and mental illness, and that more and more people are open about having mental illness should lead to a decrease in mental illness stigma. If so many people have mental health problems, having mental illness should not be ostracizing. If so many people are aware of the presence and dangers of stigmatization, they should be able to avoid stigmatizing others and themselves. Being more aware of mental disorders and symptoms should decrease stigma and challenge stereotypes about mental illness. It should be easier for people suffering from mental health problems to feel acceptance toward themselves, find support, and address their problems. Yet it is not. Despite the commonality of mental health problems, the increased awareness of mental health and illness, and the extensive resources directed toward addressing mental health problems, mental illness is still ostracizing, and mental illness stigma, negative stereotypes, prejudice, and discrimination continue to circulate within the culture.

There are two primary places where stigma persists. One is in the context of severe mental illness, like schizophrenia and bipolar disorder, especially where mental illness and homelessness intersect.[2] People who have severe mental illness that prevents them from living and working within the community, who as a result lack a home and/or a job, are commonly feared and shunned, and socially outcasted from the community. Negative stereotypes of individuals who are homeless and have mental illness abound. These include stereotypes that they are incompetent; that they are needy and dependent on others; that they are dangerous and violent; that they have an illness that is incurable and progressively worsens over time; and that they are defective human beings who have something intrinsically wrong with them. Such stereotypes are reinforced when individuals are also men and/or Black. People who have severe mental illness are typically the object of fear, avoidance, blame, shame, and disgust.

Even when people with mental illness are not homeless, they are still often the targets of negative stereotypes, prejudice, and discrimination in many areas of their lives. The more severe their illness is perceived to be, or the more blameworthy they are seen as being, the more stigma they receive.

[2] Over one-fifth (20 percent) of the population of chronically homeless people have a severe mental illness such as schizophrenia, bipolar disorder, or major depression, while about 16 percent have a substance abuse disorder (Saldua 2023).

At one time it was thought that reconceptualizing mental illness as a brain disease would decrease stigma, because it seems to shift blame for the condition away from the person, but in fact it has only increased it (Corrigan and Watson 2004; Glannon 2008; Read et al. 2006; Thachuk 2011). The view that mental illness is a brain disease suggests to many that people who have mental illness are inherently different from others, defective, unable to change or get better, and blameworthy or at least worthy of shame for having something intrinsically wrong with them (Gergel 2014; Haslam and Kvaale 2015).[3] However, the contrary psychosocial conception of mental illness as a result of trauma, family dynamics, or inadequate coping skills can be just as damaging. When mental illness is seen in this way, people are typically regarded as having a character flaw, as defective, as incompetent, as being weak or weak willed, or as immoral (Gergel 2014). Both dominant conceptions of mental illness are highly stigmatizing because they both suggest associated traits that are thought of as extremely negative.

Public stigma – stigma directed toward people with mental illness by others – persists in many places. People with mental illness experience stigma in the workforce, in housing, in healthcare settings, in educational environments, in their families, among friends, and in society at large. While many people in society are more understanding and empathetic of mental illness than ever before, in part because mental illness and mental health and well-being are talked about so much now, many other people are still stigmatizing. Sometimes this is due to a lack of knowledge about mental illness and what the lives of people who have mental illness are like; or due to a shortage of empathy and understanding; or due to the internalization and endorsement of dominant American ideals of independence, self-sufficiency, and personal responsibility that mental illness seems to violate. Whatever the cause, public stigma continues to exist despite higher rates of mental illness occurrence and despite the fact that we talk about mental illness and mental health and well-being more than ever.

The other area where stigma stubbornly persists is in the internalization of mental illness stigma by people who have mental illness themselves. People internalize mental illness stigma either when they endorse negative

[3] Haslam and Kvaale note that the brain disease model triggers essentialist thinking in which those with mental illness are thought to be radically different from others in an essentialist way that is not changeable, so that people with mental illness are seen as not being able to control their condition and cannot recover from their illness. This leads to attitudes of fear, blame, and rejection toward those with mental illness (Haslam and Kvaale 2015).

stereotypes and prejudice as applying to themselves or when their thoughts, feelings, and behavior are shaped by stigma. Such internalization occurs even when no one is actually stigmatizing the person. A person with mental illness can fear being judged on account of their illness even when no judgment actually occurs. The very worry about being judged is enough to give stigma a life of its own that infiltrates a person's consciousness and changes how they think, feel, and interact with others.

Negative stereotypes about mental illness circulate within a culture even when no one actively endorses those stereotypes. The cultural and historical contexts of mental illness stigma give stigma a power that commonality of mental disorder occurrence, awareness of mental health and mental illness, awareness of mental illness stigma, and resources directed toward treating and destigmatizing mental illness cannot erode. People are inescapably aware of mental illness stigma and its associated stereotypes and are constantly worried about being judged in relation to them. Even when no actual people hold stigmas against a person with mental illness or apply stereotypes to them, the mere fact that such stigma and stereotypes are a possibility – the mere fact that one could possibly be judged in relation to them – is enough to give this stigma and its associated stereotypes power over a person.

In twenty-first-century American culture, fear of being judged by others is common. Social media has made people, especially young people, more self-conscious than ever. This self-consciousness – awareness of how one is perceived by others – feeds a constant worry that one will be judged by others. Young people are especially prone to being anxious about other people's judgment of them, but even older people have adopted this self-consciousness about how they come across to others. Worries about other people judging oneself lead a person to be more aware of the existence of negative stereotypes that could possibly be applied to them, and more worried that such stereotypes will be applied to them. When people have mental illness, they are aware of the negative stereotypes of mental illness that circulate within a culture, and anxious about how those stereotypes can be applied to them when others perceive and judge them.

This book analyzes the process of stigmatization, both public and internalized, in the twenty-first century in Western culture, especially in the United States. Stigma is as pervasive as ever, despite greater awareness than ever of mental illness and associated stigma, and despite greater resources directed toward treating and destigmatizing mental illness. Ever since the COVID-19 pandemic in 2020–21, significant attention has been paid to mental health, yet young people and old alike are more prone to having mental health issues than ever before. The pandemic caused significant isolation and loneliness; fed unhealthy behaviors, ruminating thoughts, and dysregulated emotions; created obstacles to self-care; and caused a crisis of self-consciousness where young people especially fear being judged negatively by others. At the same time, more public attention has been paid to mental health issues in recent years, including the creation of the 988 emergency mental health crisis hotline in the United States, and more resources have been directed toward treating, understanding, and destigmatizing mental health problems. Nonetheless, such problems – and the stigma accompanying them, especially the internalized stigma that people project onto themselves – persist.

In addition, in the 2020s, homelessness has reached epidemic proportions in cities and towns across the United States (and the Western world), with unsightly encampments littering the landscape. While homelessness has many different root causes, mental illness and substance abuse are two significant factors in many people's experiences of homelessness. Mental illness stigma intersects with homelessness stigma and unemployment stigma to cause significant prejudice, shunning, avoidance, and discrimination of the people subject to these interlocking stigmas. People who are part of other minority groups, such as Black men, encounter even more public stigma due to their multiple social identities. This book assesses some of the reasons why both public stigma and internalized stigma endure in the twenty-first century, showing how negative stereotypes about mental illness are pervasive in American and Western culture regardless of many people's active disavowal of them.

Stigma and Social Norms

Mental illness stigma marks people with mental illness as bad and inferior for failing to live up to social norms that are endemic to life in the twenty-first century in a democratic, capitalist society. The American ideal of the independent, self-sufficient, rational, and autonomous person lives on throughout Western culture. This person is a rational chooser who can set goals for themselves and work toward achieving these goals through their own hard work.[4] They are self-directed, can get their own needs met,

[4] This conception of the self as a self-determined rational chooser comes from liberalism (rooted in Thomas Hobbes's and John Locke's philosophies) and more recently neoliberalism (rooted in Friedrich Hayek's economic theory) (Dean 2008, 48–49). The self-determined rational chooser is situated within a neoliberal economic framework that prioritizes privatization and commodification of goods, services, and institutions; privatization of responsibility and care; the values of freedom,

and are successful at achieving the ends they set for themselves.[5] This makes them strong, independent, and self-sufficient. They are in control of their actions, as well as their thoughts and emotions, able to self-regulate both emotions and behavior, and able to think through reasons and evidence to develop rational beliefs. As a consequence of their choices, they are personally responsible for their actions and experiences.[6] They are sane in that their perceptions of reality match the perceptions of others. They are rational in that they can calculate costs and benefits to do a reasoned analysis of what action to take; they can think through and account for consequences of different actions; they can reason from premises to conclusions; and they can weigh and account for reasons and evidence in forming and revising their beliefs. While the ideal of the rational agent may be shifting with a more nuanced understanding of human differences, those who are viewed as *less* rational and having *less* agency compared to others are still heavily denigrated.

Moreover, the ideal person is a source of safety and stability, enabling the safety of both themselves and others and stable in disposition and behavior so that they are seen as reliable and trustworthy. They are a provider to others and contribute to society through productive and efficient action. This ideal of agency and personhood emerges from Western liberal thought stemming from the Enlightenment but has morphed to support the needs of a capitalist society in the twenty-first century. This ideal is also strongly associated with an ideal of masculinity, in contrast to a feminine ideal that would support a more relational and interdependent self. The economically rational, autonomous, sane, self-controlled, self-regulated, self-directed, self-sufficient, independent, strong, safe, and stable person who is able to get their own needs met and who is responsible for their own actions is the ideal of what a human being in the Western world in the twenty-first century should be.

This conception of the self is an idealized notion in the sense that choice, control, self-regulation, self-direction, independence, self-sufficiency, action, and responsibility are viewed as ideas abstracted away from particulars, without regard to the real-world constraints on these that people

choice, and competition; and faith in the marketplace as the domain where choices and actions are made (Braedley and Luxton 2010).

5 This ideal of the self as a self-determined end-setter is rooted in Immanuel Kant's theory of rational autonomy as that which makes humans distinctly *human* (Kant 1993).

6 Thomas Lemke describes this responsibilization in terms of a neoliberal conception of the self as personally responsible for its care, esteem, worth, actions, and choice rather than these being a product of social forces where society carries some social responsibility for the outcomes of these (Lemke 2001; see also Harper and Speed 2012, 13–14).

actually face in their life experience (O'Neill 1993, 309; O'Neill 1987). These idealized conceptions of aspects of the self typify what Charles Mills calls an idealized model in ideal theory, where the ideal constitutes a moral exemplar that is seen as representing the social world without reference to real-world constraints that complicate these aspects of the self (Mills 2018; Mills 2005). Idealized models in ideal theory contrast with descriptive models of morality and politics that attempt to describe the world as it *is*, rather than as it *should be*, in relation to real-world constraints. Idealized models of the self consist of abstractions that ignore and preclude the particularities and complications of people's life experience, including limited capabilities; an intricate social environment and complicated social relationships; experiences of oppression, domination, and other injustices; real-world obligations and duties of care to others; role responsibilities; and other constraints on the expression of the self. This book on mental illness stigma philosophizes from the way stigma actually operates in the world rather than from ideal theory.

Mental illness appears to violate the social norms described above in many ways. These norm violations constitute the many negative stereotypes associated with mental illness. Such stereotypes include the view that people with mental illness are crazy, out of control, irrational, immoral, and irresponsible; that they lack sufficient agency and autonomy to be fully human; that they are damaged, weak, and unable to accomplish ends they set for themselves; that they are incompetent; that they are needy and dependent on others; that they are dangerous and violent; that they have an inherent defect; and that their illness is incurable, permanent, and worsens progressively over time. These all mark a person as bad and inferior, and lesser compared to others. They violate the social norms of what a human being should be in the Western world in the twenty-first century. When people internalize these norms, they adopt stigma against those who violate the norms, whether that is projected onto other people or onto themselves.

Stigma and the Main Themes of This Book

The first major theory of stigma comes from Erving Goffman in his seminal work *Stigma* (1963). He defines stigma as the discrediting and devaluing of a person on account of having a trait that is deemed negative to have. Stigma is a taint on a person's identity that creates a failing, a shortcoming, or a handicap on a person's social identity. Goffman identifies three areas where people may be stigmatized: their bodies (what are perceived as "physical deformities"), their character (such as having a weak will, excessive emotion, or lack of self-control), and their "tribe" – their race, nation, or religion (Goffman 1963, 3–4). Goffman locates mental illness stigma as a stigma of weak or defective character, but in our contemporary biomedical model of mental illness we might also understand mental illness as a stigma of the body. In twenty-first-century Western culture, mental illness as stigmatized can be seen as a blemish on both the body and the mind/character.

Traits such as race, gender, class, sexual orientation, ability status, mental health status, citizenship, and agency status can all be stigmatizing in certain contexts, where some identity characteristics are less valued and less credible than others. Discrediting someone involves thinking of them as having less credibility, authority, and legitimacy as a member of a community, as an autonomous agent, and even as a human being. When a person discredits another based on having a socially undesirable trait, they judge the trait to be bad and apply this judgment to the person who seems to have the trait. Devaluing a person involves seeing them as marked by badness and consequently viewing them as lesser compared to other people, having less moral standing, being less of an agent, being less of a person, or even being less than human. Stigma can therefore be understood as marking a person as both discreditable and bad (devalued) on account of possessing the stigmatized trait.

Stigma is relational; the discrediting and devaluation of people who share a certain social trait occurs as a result of viewing some people as abnormal for having the trait in relation to other people who do not have that trait and who are thus seen as normal. When normalcy is defined in part by the absence of a certain trait, the presence of that trait makes a person appear abnormal; abnormality is thus defined in relation to normalcy and involves making a normative judgment about what constitutes normalcy versus deviance. While an abnormality could be viewed as a mere difference, not inherently bad, in the context of stigma abnormality is always seen as bad. Stigma marks a person as abnormal, where the abnormality is seen as a bad thing, and thus bad to have, in relation to those who lack the stigmatized trait, who are viewed as normal and thus good.

As Goffman points out, stigmatized traits always have the *potential* to be discrediting and devaluing, and are thus *discreditable and devaluable*, making a person vulnerable to being judged negatively on account of having or seeming to have the trait (Goffman 1963, 4; Liamputtong and Rice 2022, 116–117). When people have a stigmatized trait that is not easily visible to others, they are discreditable without necessarily being discredited. Stigmatized traits are also *discrediting and devaluing*, where a person is in fact discredited and devalued when the trait is visible to others or at least appears to be present. In such a case, a person is not simply *vulnerable* to being judged by others; they *are* judged by others. The visible presence or seeming presence of a stigmatized trait invites judgment, making judgment inevitable and unavoidable.

A person does not have to possess the stigmatized trait in order to be judged; it is sufficient if they are *believed* to have the stigmatized trait, whether or not they in fact do. Judging a person as having a stigmatized trait occurs whenever the person making the judgment *believes* the person being judged has the trait or could possibly have the trait. Seeming to have a stigmatized trait or being vulnerable to being seen as having a stigmatized trait is sufficient for a person to be judged and stigmatized.

Notably, people who are or may be stigmatized hold the same beliefs about normalcy and stigma that others do. When they live in a culture that stigmatizes a certain trait, they often learn to stigmatize that trait themselves, even if they also happen to possess the trait. Goffman notes that when they are aware that they have the trait, this can result in shame and a fissure of identity (Goffman 1963, 7). The person can experience a split between what they want for themselves – to be normal – and what their self is actually like – having a seeming abnormality. This split between who they in fact are and who they want to be can cause significant problems with self-concept, self-regard, self-worth, and agency. The application of stigma to oneself can be understood as self-stigma, which is a form of internalizing stigma that I discuss below.

When a person is not aware that they have the trait, or when they do not self-identify as having the trait, they may feel distance from the trait and deny that it applies to them. Since they have internalized the stigma associated with the trait, viewing the trait as a bad thing to have, they are motivated to want to distance themselves from the trait so that it cannot be seen as applying to them. In the context of mental illness, this often involves the person denying that they have a mental illness so that the stigma associated with mental illness will not be seen as applying to them. People can go to great lengths trying to prove that they lack the stigmatized trait and engage in all kinds of rationalizations to explain away behaviors that might indicate a trait like mental illness. Stigma provides a strong motivation to deny having a stigmatized trait like mental illness.

Stigma and Power

Stigma exists in the context of power. Different components of stigma occur in the context of social, political, economic, and epistemic power relations where people who *do* the actions associated with stigma have more power of various kinds than people who are stigmatized against (Link and Phelan 2001, 375; Hipes 2016, 5). Not only does power make stigma possible but stigma also reinforces power. Stigmatizing individuals is a way of disempowering them, which perpetuates status quo relationships of power, sustaining the power and privilege of those who do the stigmatizing. Stigma thus constructs, reinforces, and reifies power relations.

Power is the ability to affect outcomes. Social power involves affecting the outcomes of other people, while political power affects the outcomes of communities, and economic power affects economic outcomes. Epistemic power is the power to affect knowledge transmission and creation and systems of meaning-making. The act of stigmatizing others reinforces and strengthens the ability of people who are not stigmatized to affect outcomes in these different contexts. This includes outcomes that bear on what those who are stigmatized can do and know, as well as how they can interact with others, what roles they can play within a community and culture, and what material and economic resources they have access to.

One way that stigma affects power relations is through the generation and perpetuation of various kinds of injustices. One form of injustice people who are stigmatized experience is epistemic injustice. Stigma undermines the ability of stigmatized people to participate in epistemic activities, including creating and transmitting knowledge, giving testimony, asking questions, sharing ideas, developing theories, and meaning-making. When a person is stigmatized, they experience what Miranda Fricker calls an identity-prejudicial credibility deficit, which is a lack of credibility afforded to people who have a social identity that is denigrated due to negative stereotypes and stigma (Fricker 2007, 30–41). This prevents them from being regarded as equal participants in epistemic practices, leading to participatory epistemic injustice, where they are viewed as insufficiently competent to make worthwhile epistemic contributions (Hookway 2010). It also leads to the testimonial injustice related to the informational perspective, where their testimony is not taken seriously and given uptake by others (Fricker 2007, 9–29; Hookway 2010). In addition, it creates hermeneutical injustice, where a person is prevented from having the hermeneutical resources to make sense of their experience (Fricker 2007, 147–174). People who are stigmatized against experience a myriad of epistemic injustices that diminish their epistemic power and their ability to participate as equals in epistemic practices.

Stigma also affects power relations in the contexts of distributive injustice and structural injustice. The discrimination that stems from stigma causes distributive injustice by preventing stigmatized people from having access to the same resources and opportunities that others have access to or that they would have access to if they were not discriminated against (Moreau 2020, 124–151; Segev 2014). This makes it so that stigmatized people often cannot get their basic needs met or meet a minimum threshold for basic functioning, violating standards of distributive justice.[7] Discrimination also creates structural injustice, where social structures are organized in a way that disproportionately burdens people who are stigmatized and puts them in a position of unfair subordination to others (Moreau 2020, 39–75; Young 2011, 43–74). Iris Marion Young notes that people experience structural injustice when they are socially situated in ways such that the confluence of individuals' and institutions' actions creates structural constraints on what they can do so that they have more obstacles that they have to overcome (Young 2011, 43–74). Stigma generates distributive and structural injustices that diminish access to resources and opportunities and restrict agency and autonomy.

Power relations benefit those who have more power in relation to others and marginalize those who have less. A person does not need to engage personally in stigmatizing acts and attitudes in order to benefit from the existence of stigma. Not being stigmatized against – being seen as normal – is sufficient for a person to benefit from the existence of stigma. Stigma reinforces the power and privilege of all people who are not objects of stigma, regardless of whether they participate in stigma, because of the way it normalizes who they are and what behavior they engage in while subjugating others.

The act of being stigmatized reduces the ability for people who are the objects of stigma to affect what people, including themselves, can do, have, and know. Thus, stigma reduces their myriad forms of power. Insofar as stigmatized people are already marginalized, stigma reinforces that marginalization, perpetuating their lessened status, lack of privilege, and decreased social, political, economic, and epistemic power. Stigma keeps the people who are objects of stigma oppressed, marginalized, and subjugated while enhancing the power of those who are seen as normal.[8]

[7] For example, John Rawls argues that primary goods that meet people's basic needs are necessary for people to have equal opportunity in a liberal society (Rawls 1971, 90–95).

[8] In addition, power affects the extent to which a person stigmatizes others. Because they have less power in society, women tend to be less stigmatizing of others than men are (Hipes 2016, 12–13).

Stigma and the Self

In addition to creating various injustices for people who are stigmatized, stigma also causes another significant harm in the ways that it shapes people's sense of self and personal identity, and the agency and autonomy that correspond to these. Stereotypes reduce people to the social trait that is stereotyped, such as mental illness, and to the stereotyped characteristic associated with the social trait, such as incompetence, dependence, or dangerousness. This erases people's particular identities and erodes the nuance and complexity of their experience.

Stereotypes cause what Elizabeth Anderson calls an "expressive harm" of shaping the mental states and behavior of both those who can be stereotyped and those who can subject others to stereotypes (Anderson 2010, 44–66). Stereotypes circulate within a culture regardless of whether individuals endorse or apply them, giving them a pervasive and persistent power of influencing people's thoughts, feelings, and behavior. Both people who can be stereotyped and those who are in a position in which they can subject people to stereotypes are obligated to acknowledge the existence of stereotypes and the ways that they color social interactions.

Moreover, stereotypes reinforce what Natalie Stoljar calls "oppressive social scripts" that determine how a person's past behavior is perceived and create expectations for how a person will act in the future (Stoljar 2015). These scripts attribute mental states and behavior to a person based on dominant conceptions of what a person who has the stereotyped trait must be like. As a result, this defines how a person's identity and agency are perceived by others, and, when internalized, forms a person's own self-concept and self-understanding of their agency and identity.

Kinds of Stigma

Theorists make several distinctions among different kinds of stigma. Public or social stigma is stigma that others apply to a person who has or seems to have a culturally devalued trait, while self-stigma is stigma that a person internalizes and applies to themselves (Conklin 2021, 17; Corrigan and Kleinlein 2005, 16–17; Thornicroft et al. 2022, 1442–1443). People project public stigma onto others when they believe or act on negative stereotypes (sometimes without endorsing them) about the people whom they are stigmatizing. Public stigma causes prejudice, where people adopt negative attitudes and have negative emotions directed at those whom they stigmatize; discrimination, where people treat some groups of people in different, unequal, and worse ways compared to others; microaggressions, where people speak and behave in subtly prejudicial ways that nonetheless cause harm; and epistemic injustice, where people treat others as less credible and as having less epistemic agency.

While individuals who are vulnerable to stereotyping are commonly victims of public stigma, prejudice, microaggressions, discrimination, and epistemic injustice, they can also stigmatize themselves by internalizing stigma. One form of internalizing stigma is self-stigma, where individuals believe that negative stereotypes about people who have a culturally devalued and undesirable trait are true and they apply those stereotypes to themselves (Corrigan and Nieweglowski 2021, 417; Veroniki et al. 2010; Thornicroft et al. 2022, 1442). When people have self-stigma, they judge themselves poorly on account of having the stigmatized trait, and this often leads to alienation, negative attitudes and emotions toward oneself, depression and anxiety, social withdrawal, and even suicidality.

Another distinction that is often made is between experienced stigma and discrimination, and anticipated stigma and discrimination (Farrelly et al. 2014; Fox and Earnshaw 2023; Thornicroft 2009). Experienced stigma or discrimination is stigma or discrimination that a person actually experiences, where they have had an experience in the past of being stigmatized or discriminated against by others. Experienced stigma and discrimination are tightly connected to public stigma, where the type of stigma or discrimination they experience is imposed on them by others, or by the public. Public stigma, or the process of others stigmatizing a person, necessarily leads to experienced stigma, where the person experiences the stigma imposed on them by others.

Anticipated stigma or discrimination is stigma or discrimination that a person believes they will experience in the future, where they expect that others will stigmatize or discriminate against them if or when others know about the person having the devalued trait. Stigma and discrimination are often anticipated when a person has had experiences of being stigmatized or discriminated against, where they expect to experience more of the same. However, it can also arise absent any actual experience of stigma or discrimination, particularly when a person internalizes stigma and is self-conscious about stigma being applied to themselves by others. Anticipated stigma can persist even when no actual public stigma exists because of the perniciousness of the internalization of stigma and people's self-consciousness about being judged by others.

People internalize stigma when they accept or are resigned to the idea that stigma can be applied to them. In the literature on stigma, theorists sometimes use "internalized stigma" and "self-stigma" interchangeably (e.g., Beltzer 2020; Fox and Earnshaw 2023), but they are actually distinct. A person can internalize stigma through at least two mechanisms. One of these mechanisms includes self-stigma, or endorsing negative stereotypes that a person applies to themselves, while the other involves seeing oneself through the eyes of others and seeing how others may apply stigma to oneself because of the way one is viewed by others. When stigma is internalized, a person sees it as applying to themselves, but this application can occur either through their own judgment of themselves or through seeing how others may judge them and either accepting or being resigned to that presumed judgment. In this way, a person can internalize stigma even when they do not believe the negative stereotypes it is based on are true and even when they do not have negative feelings (prejudice) toward themselves. They simply maintain an awareness that they may be judged, where that awareness influences their thoughts, feelings, and actions.

This latter form of internalized stigma is closely related to anticipated stigma. When a person is aware that they may be judged according to negative stereotypes, they may form expectations that others will indeed judge them in the future. In this way, they may anticipate being stigmatized against. However, the two terms are separable. A person can be aware of the possibility of being judged without necessarily *expecting* others to do so. To distinguish these two ways of internalizing stigma, I will call the application of stereotypes and prejudice to oneself "self-stigma" and refer to the acceptance or resignation that others may apply stereotypes and prejudice to oneself, where this awareness informs their mental states and behavior, as "attendant stigma," in that the person has no choice but to pay attention to the fact that they have the potential to be judged and stigmatized by others. For both kinds of internalized stigma, the stigma is integrated into and guides one's thoughts, feelings, actions, and even identity.

All of these forms of stigma – including experienced/public stigma and discrimination, anticipated stigma and discrimination, and internalized stigma – cause a multitude of harms and problems for people, including people who are stigmatized on account of having mental illness. Experienced/public stigma and discrimination are associated with decreased housing stability and increased homelessness, less access to opportunities such as education and resources such as loans, more challenges with recovery, low quality of life, and reduced community functioning (Beltzer 2020, 679; Mejia-Lancheros et al. 2021). Internalized stigma often decreases hope, self-esteem, empowerment, and quality of life. In addition, internalized stigma can result in negative attitudes about treatment, less willingness to seek treatment, and decreased treatment adherence. This impairs a person's ability to receive effective treatment, helps perpetuate symptoms, and thus reinforces public stigma. Anticipated stigma can lead to internalized stigma and causes psychological distress and negative physical health (Beltzer 2020, 679). It also often causes individuals to self-limit activities, reducing their options and creating barriers to participation, thus limiting agency and autonomy. Many of these harms are explored in this book.

Components of Stigma

Bruce G. Link and Jo C. Phelan identify several components of stigma: labeling, stereotyping, separation (or moral distancing), status loss (including dehumanization), and discrimination (Link and Phelan 2001, 367). I would add three additional components to this list: prejudice, which is the affective response one has to people who are labeled as different and viewed as inferior because of that difference; microaggressions, which are the subtle speech and behavior perpetrators commit that cause harm to those who are targeted; and epistemic injustice, which is the injustice of not being taken seriously as a knower. Many of these components of stigma apply to internalized stigma as well, where people self-label, apply stereotypes to themselves, feel morally distanced from others, view themselves as lesser (often resulting in lowered self-worth), and self-limit what activities and relationships they feel they have a right to participate in or be part of. Here we will look at these components of stigma more closely in the context of public stigma, where others stigmatize an individual for having a culturally undesirable social trait. In Chapter 1, I examine these components of stigma both in the specific context of mental illness and in the context of internalized stigma.

Labeling and Stereotyping

Stigma begins with labeling, when people identify and label differences between various groups of people, noticing and naming different traits that groups of people share (Link and Phelan 2001, 367–368). These differences are understood in a cultural context that views certain traits as undesirable, leading to the adoption of negative stereotypes of people who have these traits and the application of these stereotypes to specific individuals. Because stereotypes circulate within a culture independently of individuals' endorsement of them, labeling someone as a member of a social group that shares a culturally undesirable social trait (such as having mental illness) automatically invokes negative stereotypes. When people are seen as being a member of a group that shares a culturally undesirable social trait, they make automatic associations between that social trait and cultural generalizations that are made about people who have that trait, and apply those associations to specific individuals who have that trait in the process of stereotyping. Labeling and stereotyping are thus inherently linked.

Theorists writing on stereotype content identify two dimensions that are typical in the content of stereotypes: warmth and competence. Warmth has to do with a person's sociability, friendliness, good-naturedness, trustworthiness, sincerity, and morality, while competence involves a person's capability, assertiveness, agency, intelligence, skill, efficacy, independence, confidence, achievement, prestige, and competitiveness (Boysen 2017, 100; Canton, Hedley, and Spoor 2023, 481; Fiske 2018, 67; Follmer and Jones 2017, 493).[9] People who are seen as "normal" and as fitting the various social norms described above – such as those with adequate rational agency; and those who are American, middle class, and/or White – are regarded as having both high competence and high warmth (Fiske 2018, 68). This marks them as normal and acceptable, and attitudes toward them are usually warm, welcoming, respectful, and even worthy of admiration.

Many stereotypes emphasize one dimension as high and the other as low, giving rise to mixed emotional reactions that value and denigrate people at the same time. When people are viewed as having high warmth and low competence – such as the elderly, the disabled, young children, and certain ethnic groups like Mexicans, Latinos, and Africans – they are typically pitied and sympathized with, leading to paternalistic expressions and actions. When people are viewed as having low warmth and high competence – such as rich people and business people, and certain ethnic groups like Asians, Germans, and Jews – they are often regarded with respect and envy, but also distrust, resentment, and even fear (Cuddy, Fiske, and Glick 2002; Fiske 2018, 68). Fear results because lack of warmth is sometimes tied to a perception of dangerousness (Sadler, Meagor, and

9 Note that competence is defined here in a broad way as encompassing various attributes of capability, agency, and effectiveness. This is different from how competence is defined in legal contexts in relation to people with mental illness, where the term refers to decision-making capacity (Buchanan and Brock 1989, 18; Halpern 2001, 103). In general terms, competence refers to the ability to be successful at tasks (Edwards 2010, 275–276). In the context of stereotyping people with mental illness, competence refers to how a person functions in various life domains, including social interactions, social relationships, work, school, civic life, and leisure activities. In this book, I use the term competence to refer to general functioning in daily life activities, except where discussing the stereotype content model.

Kaye 2012, 916). People who are seen as having both low warmth and low competence – such as homeless individuals, drug addicts, refugees, and undocumented immigrants – are regarded as the lowest of the low, leading to attitudes of disgust and contempt (Fiske 2018, 68). People with mental illness can fit any of these categories depending on what mental illness they have or are seen to have and depending on what other social identities they have and how the stigma associated with these intersects with mental illness stigma.

Prejudice

Prejudice, sometimes used synonymously with the term "bias," is a negative attitude toward people who share a certain group membership that is viewed unfavorably because of negative preconceived notions about members of that group (Allport 1958, 8). It can also be understood as thinking poorly of certain people based on their group membership without warrant, or without having a morally justifiable reason (Allport 1958, 7; Hehman and Neel 2024, 1237; Katsekas and Lemay 1996, 2). Prejudice can be operationalized through two judgments people make, based on what they think of the group they are judging (the "out-group") and how this compares to what they think of the social group that they themselves belong to (the "in-group") (Hehman and Neel 2024, 1237; Allport 1958).

Prejudice is primarily affective, involving emotions, reactive attitudes, and unconscious/automatic associations (Corrigan and Kleinlein 2005, 16–17). Some of the emotions prejudice invokes are fear, anger, disgust, and contempt. Prejudice is also manifested through bodily expressions such as verbal tone, word choice, body comportment, gestures, and body language. How people are socialized, what family values they grow up with, what religious beliefs they hold, what political affiliations they align themselves with, and what educational experiences they have all influence what prejudices people have (Katsekas and Lemay 1996, 3).

People form prejudice as a result of judging others as inferior. This judgment of inferiority leads people to develop automatic negative associations of those who share the stigmatized trait associated with the social group that one is prejudiced against, causing negative emotions like fear, anger, disgust, and contempt, and negative reactive attitudes like blaming and shaming. Prejudice contributes to behavioral responses of shunning and avoiding people who have the undesirable trait and treating them differently and negatively compared to other people, creating discrimination. Prejudice is the affective aspect of stigma that links the cognitive aspect of stereotyping and the evaluative aspect of judging people with the behavioral aspect of discrimination and microaggressions.

While prejudice can be knowingly held and endorsed, it is often held unconsciously and unknowingly through implicit bias. Implicit bias is a negative attitude that a person holds toward those who share a socially undesirable and negatively stereotyped social trait, usually held unconsciously and unreflectively (Holroyd 2012; Holroyd, Scaife, and Stafford 2017; Levy 2017). Despite being outside the person's awareness, implicit bias colors how a person feels and thinks about, and interacts with, those against whom they are biased. As a result, they hold negative affective reactions to the person – in other words, prejudice – including negative emotions like fear, anger, disgust, and contempt, and negative reactive attitudes like shaming and blaming. This affects their behavior toward the person so that they are more avoidant of, judgmental of, aggressive toward, or punishing toward the person, or have other negative action tendencies toward the individual. Prejudice and implicit bias influence people's perceptions of others so strongly that some philosophers are skeptical that we can ever have accurate understanding of those against whom we hold biases, or that we can ever avoid epistemic and other injustices toward those against whom we are biased (Pasnau 2022; Saul 2013).

Moral Distancing and Outcasting/Social Exclusion

People who have culturally undesirable traits and are labeled based on these traits are viewed as separate and fundamentally different from those who do not have these traits, creating a separation of "us" versus "them" (Link and Phelan 2001, 368–370). This creates moral distance between those who do not possess a culturally undesirable trait and those who do. This moral distancing often serves to "other" the people who are seen as different, so that they are viewed as abnormal or having a problem or defect on account of that difference. Separation and moral distancing are often self-reinforcing, as "other"-ing people who appear different from oneself can reify one's own status as being unlike "them" so that one's own status is taken to be "normal" and correct while "theirs" is taken to be abnormal and wrong. The more that people are seen as different from others, the more distant they seem to be from what is taken to be "normal."

Separation or moral distancing often results in outcasting and social exclusion. Outcasting a person involves seeing them as someone who does not belong to the community and rejecting them and their ability to participate in social, moral, and epistemic practices of the community. Social exclusion often arises through shunning and avoidance by others. It can also occur when the person with the culturally undesirable trait outcasts themselves in anticipation of being outcasted by others and withdraws socially from the community. Outcasting and social exclusion can diminish agency and autonomy, create loneliness and stress, cause or exacerbate mental disorder symptoms, and decrease quality of life.

Status Loss and Dehumanization

Regarding some people's differences as abnormalities and marking them as bad for having these perceived abnormalities grants these individuals a loss of status, where they are seen as inferior to others in various ways (Link and Phelan 2001, 370–371). Some of the ways people can be seen as inferior to others include being perceived as less capable, less resourced, less resourceful, less intelligent, less rational, less of an autonomous agent, less moral, less having a unique identity, and even less than human. Having a diminished status can seem to others to justify trying to avoid interacting with those who have lesser status, resulting in social exclusion, or to justify different, worse, and unfair treatment of those with lesser status, resulting in microaggressions, discrimination, and epistemic injustices.

Status loss often results in dehumanization, which is the degradation of a person's humanity. Dehumanization can be understood broadly as a setback of significant human interests (Mikkola 2016, 145 and 164) or more narrowly as treating a person as subhuman, such as by comparing them to a monster, animal, machine, or trickster.[10] The less warmth and competence a person is perceived to have, the more they are dehumanized (Fiske 2018, 69; Kuljian and Hohman 2023). Warmth and competence both correspond with having characteristically human qualities such as rationality, self-control, agency, autonomy, compassion, benevolence, morality, unique identity, and personhood.

In the context of mental illness stigma, it is sufficient to understand dehumanization as treating people with mental illness as subhuman. Here

[10] David Livingstone Smith describes dehumanization as thinking of someone as subhuman by comparing them to a monster or animal (Smith 2016; Smith 2014). In the social psychology literature, dehumanization is distinguished between animalistic dehumanization, which denies a person as having uniquely human characteristics such as having civility, morality, and sophisticated thinking, comparing them to animals that act on instinctive impulses; and mechanistic dehumanization, which views a person as cold, rigid, inert, and lacking agency, comparing them to inanimate objects, machines, and automatic robots (Haslam 2006; Kuljian and Hohman 2023). The idea that humans can be tricksters as another way of being perceived as subhuman is my own.

they are viewed as being lesser than other people on account of seeming to have less competence, rationality, sanity, self-control, agency, autonomy, morality, unique identity, and personhood compared to others due to their mental illness. People with mental illness are commonly subject to dehumanization due to the perception that they lack various characteristically human abilities and attributes such as these (Boysen, Chicosky, and Delmore 2023; Boysen et al. 2020). Being viewed as deficient in these ways can make a person seem more like a monster, an animal, a machine, or a trickster, rather than a human being. These ways of viewing a person as subhuman can seem (in some people's eyes) to justify worse and unfair treatment, including microaggressions, social exclusion, discrimination, and epistemic injustice, as a result.

Microaggressions

Microaggressions are subtle displays of stigma through speech or behavior that demeans a person in a way that often does not get noticed by perpetrators or bystanders. Microaggressions are a form of behavioral discrimination constituting speech or behavior that treats someone differently from and worse than others. Types of microaggressions include microassaults (speech or behavior that is conscious and intended to cause harm), microinsults (speech or behavior that demeans a person's marginalized identity in a subtle way that often goes unnoticed), and microinvalidation (speech or behavior that discounts or dismisses a person's thoughts, feelings, and experience) (Sue et al. 2008, 330–331; Sue et al. 2007, 274). The invisibility of microaggressions is part of what perpetuates them, as perpetrators are often well-intended people who do not recognize that they are acting in unconscious ways that cause harm (Fleras 2016, 9–10; Sue et al. 2007, 275). This can cause people to adopt a defensive attitude when they are called out on their microaggressions (Friedlaender 2018, 13) and contribute to what Jean Harvey calls "civilized oppression," or daily acts of oppression that are subtle and often unconscious yet persist over time as they perpetuate status quo power relations (Harvey 2007, 31–33).

While perpetrators often do not realize they are committing microaggressions (particularly microinsults or microinvalidation), those who are targeted by microaggressions are often all too aware of their presence and their power to harm. Some of the harms they create include feelings of powerlessness, feelings of invisibility, forced compliance and loss of integrity, and pressure to represent one's group (Sue, Capodilupo, and Holder 2008, 333–334). They both perpetuate larger structural inequities (Friedlaender 2018, 7–8) and promote "everyday suffering" (Fleras 2016, 9). Microaggressions create cumulative harm, where a single discrete act may not make much of a difference, but being subject to microaggressive behavior over time has a significant impact (Friedlaender 2018). Sometimes perpetrators of microaggressions believe that people who are targeted are overreacting to their single, discrete action because they are unaware of the systemic and sustained effects of multiple microaggressions committed over time.

Because of their invisibility, and because perpetrators of microaggressions often deny that they are committing these actions, those who are targeted by microaggressions sometimes question themselves, wondering if they really experienced what they thought they experienced. In this way, they second-guess and doubt themselves (Sue et al. 2007, 279). This self-doubt further diminishes their power and status, as the perspective of perpetrators who deny they did anything wrong gets privileged over the perspective of those who are targeted, who second-guess what they thought they experienced (Fleras 2016, 3). Through this, microaggressions reinforce existing power relationships that privilege those who act in ways that harm already marginalized people who are subject to both conscious and unconscious stigma.

Discrimination and Injustice

Moral distancing, status loss, and prejudice often result in discrimination, where people who have the culturally devalued trait and who are labeled as different and lesser than others are treated not only differently and unequally compared to others but also negatively (Link and Phelan 2001, 372–373). Discrimination is different, worse, and unfair treatment of certain groups of people based on their possessing a culturally devalued trait. This treatment can be understood in many different ways. Discrimination can involve treating certain groups of people in ways that make them worse off than they would be without that treatment (Lippet-Rassmussen 2014a). It can be seen as a distributive injustice in which certain groups of people receive an unfair distribution of resources and opportunities (Segev 2014) or as a distributive injustice in which certain groups of people receive insufficient basic goods to be able to be equal participants in society (Moreau 2020, 124–151). Discrimination can also be seen as a violation of dignity in which certain groups of people are demeaned (Hellman 2008, 6). It can be viewed as a structural injustice involving unfair subordination (Moreau 2020, 39–75). And it can be regarded as a loss of deliberative freedom (Moreau 2020, 77–119) or as an impingement of autonomy.

Discrimination is itself an injustice, but it is also deeply connected to other types of injustice, especially distributive and structural injustices. Discrimination often results in material, economic harms of allocating access to resources and opportunities in unequal and unfair ways. This breaches standards of distributive justice. Discrimination also tends to involve subjugating some people to the power and authority of other people, based on how they are socially situated within unjust social structures that benefit some people at the expense of burdening others. This violates standards of structural justice.

Distributive justice demands that all people have at least a certain threshold of necessary goods where their basic needs are met and they have the goods necessary to be able to achieve the outcomes of a just society. In a welfarist conception of justice, this means having a sufficient amount of goods to be able to satisfy one's reasonable preferences and live lives of equal desirability (Arneson 1989; Dworkin 1981a). In a liberal conception of justice, this means having enough resources to overcome handicaps and risks to live a life of one's choosing based on one's own conception of the good (Dworkin 1981b). This involves having sufficient primary goods to be able to have equal opportunities to hold various social positions and offices in a Rawlsian conception of justice (Rawls 1971, 90–95) or to be able to inhabit various social roles (Reader 2007, 55). In a needs-based theory of justice, this requires having the array of basic goods necessary to have agency and autonomy (Brock 2005; Copp 1992), or to be able to have one's human rights fulfilled (Kelly 2004). There are a multitude of ways distributive justice can be understood, but all involve having fair outcomes, where people have access to the goods they require to live minimally decent and ideally worthwhile lives (Pogge 2008, 43–44).

Distributive injustice is the injustice a person faces when resources, opportunities, and support are distributed among people in an unfair way that makes certain people unable to access an adequate amount of these goods sufficient to meet their needs. With distributive injustice, people are unable to get their basic needs met and unable to achieve the demands of justice, whether that be preference satisfaction; living a life of equal desirability; living a life based on one's own conception of the good; having the opportunity to hold social roles, positions, and offices; having agency and autonomy; or having one's human rights fulfilled. This denies people the ability to live a worthwhile or even minimally decent life.

Discrimination also often involves structural injustice in which social structures unfairly and disproportionately put more constraints on the actions of certain groups of people than others. This unfairly subordinates people who are more burdened by social structures to those who benefit from these structures and treats those who benefit from structures as having more moral worth than those who are burdened by them. When social structures harm some while benefitting others in systematic ways based on how people are socially situated within the relevant social structures, this violates the moral principle that structural benefits and burdens should be distributed fairly, treating all people equally, so that all people have sufficient options for action.

Structural injustice is the injustice a person experiences as constraints on their agency due to the way social structures operate. While distributive justice concerns having just outcomes, structural justice involves the justness of the social structures and processes that lead to these outcomes. Social structures consist of the way individuals, institutions, and systems are positioned in relation to each other in society to enable and constrain the actions of individual and institutional agents in different ways. As Iris Marion Young notes, individual and institutional agents occupy various social-structural positions that give them different ranges of opportunities and limitations on their action (Young 2011, 45). People experience structural injustices when they are positioned in ways that subject them to systematic burdens and harms based on how they are situated socially, while other people are positioned in ways that allow them to experience systematic benefits (Young 2011, 52). People who are vulnerable to various social inequities and harms, including those with mental illness, are subject to more systematic burdens as they face more limitations and experience more constraints on their action due to how they are socially situated. Systemic discrimination sets limitations on vulnerable people's actions, in part by preventing them from having access to the basic goods they need to function in society, so they have less agency.

In addition to putting more limitations and obstacles on some people's actions compared to others', unjust social structures unfairly subordinate some groups of people to others based on a perception that the subordinated group has less moral worth than nonsubordinated groups (Moreau 2020, 124–151). This unfair subordination results in those groups with perceived greater moral worth having power and authority over those groups with perceived lesser moral worth. This power and authority allows them to create systems within social structures that benefit them, and that encourage those with less power and authority to collude in them, thinking these systems benefit them as well, even though they do not. Discrimination maintains status quo power relations and can lead those who are discriminated against to feel resigned to their lot in life, believing that their situation is inevitable or even good for them.

The negative treatment that people who are discriminated against typically experience consists of denying or preventing them from being able to access resources and opportunities that others are able to access, or that they would be able to access if they were not treated in this way. Employment, housing, education, and healthcare are some of the resources and opportunities that are available to many but that certain people may be denied on account of their having or being seen to have a culturally undesirable trait that is the basis for discrimination. Discrimination can occur in many different areas of a person's life and can lead to social outcasting and shunning, decreased capabilities, diminished agency and autonomy, and poorer quality of life.

Epistemic Injustice

People who are stigmatized against experience various forms of injustice, including the distributive and structural injustices that stem from discrimination. A distinct form of injustice that stigmatized people face is epistemic injustice, or injustices in their capacity as a knower. Epistemic injustice stems from and is related to all of the other components of stigma, including labeling, stereotyping, prejudice, moral distancing, social exclusion, status loss, dehumanization, microaggressions, and discrimination. As already noted, the two primary forms of epistemic injustice, originally identified by Miranda Fricker, are testimonial injustice and hermeneutical injustice (Fricker 2007). In testimonial injustice, a person's testimony is not taken seriously due to prejudice against their social identity, while in hermeneutical injustice, a person is denied the hermeneutical resources that allow them to make sense of their experience, again due to prejudice against their social identity.

Epistemic injustice can be transactional, institutional, or structural. Transactional epistemic injustice occurs when individual agents interact with each other in ways that prevent certain groups of people from being seen as equal participants in epistemic practices. Institutional epistemic injustice consists of institutions preventing certain groups of people from being able to participate in epistemic activities through policies and practices that preclude them from being recognized as credible epistemic participants.[11]

[11] Elizabeth Anderson notes that social institutions can be epistemically unjust (Anderson 2012). She conflates institutional epistemic injustice with structural epistemic injustice, but these can be distinguished from each other.

Structural epistemic injustice exists when social structures unfairly and disproportionately burden some groups of people by putting unintentional yet systematic epistemic constraints on their epistemic behavior. People who are stigmatized against can experience epistemic injustices at all of these levels.

As we see, stigma involves several mental and behavioral processes (McCullock and Scrivano 2023). Marking people as different, labeling them, and forming stereotypes about people based on traits that they share are cognitive processes. Judging people for being different and assessing them as inferior are evaluative processes. Prejudice against people who share the stigmatized trait, where one has negative associations, emotions, and reactive attitudes toward stigmatized people, is an affective process. Treating stigmatized people as different, in unequal and negative ways, and thereby discriminating against them or committing microaggressions against them, is a behavioral process. Other behavioral processes involve subjecting people to distributive, structural, and epistemic injustices. These different components of stigma connect with each other to create an experience of devaluing a person on account of them having a culturally undesirable trait such as having mental illness. In Chapter 1, I examine what these components of stigma look like specifically for people who have mental illness.

The Contours of This Book

The general aim of this book is to examine what mental illness stigma consists of, and to develop an account of what makes it harmful and thus wrongful, as well as to use this account of resultant harms to make a positive proposal of how mental illness stigma can and should be addressed. In making this analysis, the book explores several aspects of mental illness stigma, including the process of stereotyping, the expressive harm of stereotypes, the role of social norms embedded within stereotypes in creating adaptive preferences and shaping behavior, the moral distancing and status loss involved with social exclusion and dehumanization, and the harm of discrimination. The book shows how mental illness stigma creates a variety of harms for those who are stigmatized against, in part because it has a cultural power that persists despite more attention and resources than ever being directed toward mental illness treatment and destigmatization. As a result, addressing stigma requires lessening its cultural power and changing the social norms that mental illness is perceived as violating.

The main argument in this book is that mental illness stigma has lasting power to control the thoughts, feelings, behavior, and social interactions of people in ways that gravely harm those who are stigmatized; as a result, it is paramount that mental illness stigma be challenged. I start by analyzing what stigma is, drawing on two leading models of stigma, and argue that both public stigma (stigma that is imposed on people with mental illness by others) and internalized stigma (stigma that is projected by the person with mental illness onto themselves) are pervasive in a mental illness context. The book is organized around several aspects of mental illness stigma that it focuses on: the process of stereotyping and the kinds of judgments made while stereotyping; the way stereotypes cause harm to self and identity (primarily through internalizing oppressive social scripts and stereotype threat); the role of social norms embedded within stereotypes in creating adaptive preferences that unduly restrict options and damage self-concept; threats to the relational self, including the moral distancing and status loss involved with social exclusion and dehumanization, and the precariousness of identity and autonomous agency involved with belonging uncertainty; the diminishment of epistemic credibility and agency; and the various material and agential/identity harms and distributive and structural injustices that result from discrimination. Public stigma creates some of these harms, while internalized stigma causes others (and some harms result from both).

I situate my project in relation to the philosophical literature on stereotyping, prejudice, discrimination, dehumanization, relational autonomy, and adaptive preferences, drawing on many philosophical concepts and theories in these areas to develop my own account of mental illness stigma. I also engage with the social psychology and sociology research on the experience and perpetuation of mental illness stigma. With this engagement, I aim to offer a robust philosophical account of what mental illness stigma consists of, how it harms people who are stigmatized against, and how it can and ought to be challenged.

Scope

The scope of this book covers people who have or who are regarded as having mental illness in twenty-first-century Western culture, and the general public who interact with them. The book analyzes the process of stigmatization in the twenty-first century in Western culture, especially but not exclusively in the United States. It looks at how people who have or are seen as having mental illness are viewed and treated by others and how stigma colors this perception and treatment. The book also examines how people who have mental illness regard and behave toward themselves, and the role that internalized stigma plays in that treatment. Social norms of twenty-first-century Western culture greatly affect how people with mental illness or who are presumed to have mental illness are seen and treated, and cultural stereotypes that persist even as cultural attitudes have changed influence how the public interacts with people with mental illness. This book examines the social norms, cultural trends, and persisting stereotypes that affect the treatment of people with mental illness in twenty-first-century Western culture.

Methodology

This book proceeds through analytical argumentation to offer a framework for understanding stigma, particularly mental illness stigma, and for accounting for the harms that stigma causes to people who have mental illness, as well as to develop a positive proposal for how stigma can and should be challenged. Philosophical theories of stigma, stereotyping, dehumanization, discrimination, autonomy, and adaptive preferences are explained and examined in the process of putting forth an account of mental illness stigma. While the book mainly engages with philosophical literature related to stigma and its harms, it also reviews psychology and sociology research on mental illness stigma to give empirical information about how stigma is experienced and perpetuated. The book is written in a clear analytical style that explains terminology and theory in an accessible way. The chief project of the book is to analyze stigma, assess its harms, and provide a positive proposal for what can be done to reduce stigma.

Chapter Outline

The book proceeds as follows. Chapter 1 starts by analyzing the components of mental illness stigma to show how people with mental illness experience stigma in their daily lives. Drawing from Link and Phelan's model of stigma (Link and Phelan 2001, 367) and Corrigan and Kleinlein's model of stigma (Corrigan and Kleinlein 2005, 16–17), I develop my own account of the components of mental illness stigma. These include labeling, stereotyping, prejudice, moral distancing, social exclusion, status loss, dehumanization, microaggressions, discrimination, and epistemic injustice. Using stigma theory and empirical evidence from the social psychology literature, I show how these components of stigma manifest in (public) mental illness stigma and in internalized stigma. The chapter proceeds with an overview of the social psychology literature that demonstrates some of the factors that impact mental illness stigma, including beliefs, political values, cultural values, socioeconomic status, education, and gender. The chapter concludes with an examination of how many people experience compounding stigmas when they are subject to other kinds of stigma besides mental illness stigma.

Chapter 2 assesses what stereotypes are and explains what makes them both wrongful and harmful. The chapter begins by defining stereotypes, explaining their relationship to prejudice and implicit bias, and showing how they are maintained due to cognitive biases. Three sorts of judgments are made in the process of stereotyping, including a judgment about whether a stereotype is true or not, a judgment about whether a person conforms to the stereotype, and a judgment about whether a person is good or bad for fitting the stereotype; various factors go into making these judgments. What makes stereotypes wrongful is their rigidity, their frequent falsity, and the way they overgeneralize about a person's experience so as to erase its nuance and complexity. The chapter ends by looking at descriptive and normative components of stereotypes and showing that negative stereotypes always make a normative judgment about the badness and inferiority of a person who fits the stereotype.

Chapter 3 analyzes some of the ways that stereotypes harm people's sense of self and identity. One way is through what Elizabeth Anderson calls "expressive harm," which is the harm that results from the unwitting and inevitable perpetuation of stereotypes regardless of whether people endorse these stereotypes (Anderson 2010, 44–66). When someone interacts with a person who has a stereotyped social trait, the negative stereotype associated with that trait always hovers in the atmosphere and controls people's thoughts, feelings, behavior, and social interactions, even when people actively disavow the stereotype. Other ways that stereotypes harm people's sense of self and identity are through the internalization of what Natalie Stoljar calls "oppressive social scripts" (Stoljar 2015) and through the phenomenon of stereotype threat. In addition, expressive harm, internalization of oppressive social scripts, and stereotype threat all diminish a person's autonomy. Through these mechanisms, people with mental illness incorporate negative stereotypes into aspects of their experience and identity in harmful ways.

Chapter 4 shows how internalized stigma often results in adaptive preferences that harm a person. When people incorporate aspects of negative stereotypes into their identity, they sometimes develop adaptive preferences by internalizing harmful social norms and beliefs embedded within these stereotypes. I show how people with mental illness often develop goals and desires that are shaped by these beliefs and social norms, which limits what they believe they are capable of, thus reducing their options for action and truncating their agency and autonomy. While adapting desires to one's circumstances can be positive, as in positive adaptation, it is negative when it is harmful to a person. The adaptive preferences that result from this can be seen as rationality deficits that are oppressive and nonautonomous and that damage well-being and flourishing. Internalizing harmful social norms that are nested within negative stereotypes unduly restricts a person's life in a way that inhibits their flourishing and prevents them from exercising full autonomous agency.

Chapter 5 assesses harms that people with mental illness experience which are related to how their self is constituted, in particular harms of social exclusion and dehumanization that result from status loss and moral distancing. Having a sense of belonging and being accepted as an equal member of a moral/epistemic/social community are important parts of being viewed as (and viewing oneself as) a full human being; these are also critical for developing and exercising autonomous agency as well as for well-being and flourishing. People with mental illness are often excluded from these communities as a result of public stigma, shunning, avoidance, social exclusion, microaggressions, and discrimination. Through the process of stigmatization, they are often marked as bad and inferior and viewed as lesser compared to others, which dehumanizes them. This chapter shows how a sense of belonging and being accepted as an equal member of a community is essential for the expression of the relational self, as well as for being an autonomous agent, and how the status loss and moral distancing of both public stigma and internalized stigma threaten these. Both outright social rejection and belonging uncertainty are examined in this context.

Chapter 6 examines what makes moral and behavioral discrimination (expressions of public stigma) wrongful. Discrimination, as I am using the concept, involves differential treatment where some people are treated in different, unequal, and worse ways compared to others, and where that differential treatment is based on possessing a socially undesirable trait that marks a person as bad and inferior. Discrimination is wrongful because it harms people in a variety of ways, impacting their circumstances, resources and opportunities, options, agency, autonomy, and well-being. Discrimination causes material disadvantage, and it constitutes a distributive injustice that denies people access to resources and opportunities others have access to, for no morally salient reason (a comparative distributive injustice), as well as a distributive injustice that prevents people from having the basic goods necessary for equal participation in society (a baseline distributive injustice). It also demeans a person by degrading them; it unfairly subordinates a person through social structures that disregard their needs and interests; and it causes agential harms, including a loss of deliberative freedom and decreased autonomy. This chapter reviews the philosophical literature on discrimination to provide a pluralistic account of the many harms discrimination causes to people with mental illness, which altogether make discrimination wrongful.

After several chapters explaining the harms and wrongfulness caused by stigmatization of mental illness, including both public stigma and internalized stigma, Chapter 7 looks at various sorts of interventions that can be implemented to address stigma. These include individual actions that the person who is stigmatized can engage in to help them cope with or resist stigma, and actions that other people are obligated to perform in order to decrease stigma they may endorse or perpetuate unwittingly. These also include structural changes that social institutions and systems must undergo to make social structures less stigmatizing and more supportive of people with mental illness, and social and cultural interventions that increase the belongingness and acceptance of people with mental illness into the community as well as transform social norms to be more supportive of people with mental illness. In addition to using philosophical argumentation, this chapter draws on empirical literature in social psychology that examines what works to reduce and resist stigma.

Mental illness stigma is pervasive despite great strides we have made toward increasing awareness of mental health problems and the stigma that has always accompanied them. In examining what mental illness stigma consists of and how it is harmful to those who are stigmatized against, this book explores why it is so widespread and how it can be addressed. Although deeply rooted within our culture, mental illness stigma can be resisted and challenged. But first we need to understand it. To do that, let us turn to Chapter 1.

CHAPTER 1

An Overview of Mental Illness Stigma

Introduction

People with mental illness experience stigma in specific ways. They are subject to negative stereotypes applied to themselves by others through public stigma or by themselves through internalized stigma. They are seen as different and abnormal, and marked as bad and lesser compared to other people, leading to moral distancing and social exclusion, as well as status loss and dehumanization. They are subject to the negative emotions and reactive attitudes involved with prejudice, as well as to microaggressions, which are subtle behaviors that others commit against them largely due to implicit bias. They experience distributive and structural injustices as well as other harms due to discrimination, and they are also subject to an array of epistemic injustices.

Stigma is projected onto people with mental illness not only through the public stigma from others but also through internalized stigma from themselves. Internalized stigma can consist of the self-stigma of applying negative stereotypes to themselves or the attendant stigma of seeing themselves through the eyes of others, where they are aware that they may be judged by others according to such stereotypes and have to respond to that stigma in their interactions with others. When people with mental illness internalize stigma, they impose the various components of stigma onto themselves.

There is a hierarchy of stigma regarding different disabilities and various life difficulties, and mental illness stigma is at the top. One study showed that people stigmatize those with mental illness more than those who have physical illness, receive welfare, or have no mental illness in general (Robb and Stone 2016). Also, some mental illnesses are more stigmatized than others. People have more prejudice toward those with schizophrenia than they do toward those who have depression (Bizumic, Gunningham, and Christensen 2022) and even bipolar disorder (Burguet and Girard 2024).

Stereotypes mediate the degree to which a mental illness is stigmatized. Mental illnesses that are thought to be more severe, more permanent/ incurable, more dangerous, and more blameworthy are generally more stigmatized than those thought to be less severe, more temporary, safer, and less blameworthy.

Various factors contribute to mental illness stigma. These include beliefs, political values, cultural values, socioeconomic status, education, and gender. Many people who have mental illness also have other minoritized social identities and experience multiple kinds of stigma concurrently directed at their various social identities. This compounded stigma can be especially difficult to resist and overcome, and can make it difficult for a person to seek treatment for their mental illness. Mental illness stigma is pervasive and pernicious in many ways.

We will start this chapter with an examination of how mental illness stigma manifests in the various components of stigma.

Components of Mental Illness Stigma

Mental illness stigma consists of several components: labeling people who appear to have mental illness as having mental illness; stereotyping people based on traits they share on account of seeming to have mental illness; separating people with presumed mental illness from those who do not seem to have mental illness and are consequently regarded as "normal"; status loss of people with presumed mental illness, including diminished attributed rationality, self-control, morality, responsibility, agency, autonomy, personal identity, and personhood, as well as dehumanization; microaggressions committed against people with mental illness; discrimination against people with mental illness who are excluded from participating in various activities that people normally participate in; and epistemic injustices that undermine their capacity as a knower. Mental illness stigma also involves directing reactive attitudes such as fear, disgust, and contempt toward people with presumed mental illness and subjecting them to blame and shame for both their behavior and for simply having a mental illness.

People with mental illness are subject to these various forms of stigma both through the public stigma that is projected onto them by others and through the internalized stigma they impose on themselves. This includes the self-stigma of applying stereotypes to oneself and the attendant stigma of seeing oneself through the eyes of others and internalizing how others may judge oneself according to stereotypes. Here I examine how these various components of stigma manifest through both public and internalized stigma.

Labeling

Labeling occurs when other people identify and label someone as having mental illness. Sometimes labeling occurs when a person has actual knowledge of another's mental illness, and sometimes it occurs when the person believes another is mentally ill based on cues they express, such as certain behaviors, expressed thoughts or emotions, or appearance. Studies have found that diagnostic labeling (labeling someone as having a specific mental disorder) increases stigma of heavily stigmatized disorders such as schizophrenia and addiction but does not affect stigma of less stigmatized conditions such as depression or autism (O'Connor et al. 2022). However, labeling is the first step of expressing all forms of stigma and invites stereotyping, social distancing, prejudice, microaggressions, discrimination, and epistemic injustices.

Being labeled as having a mental illness can be empowering for some people and disempowering for others. For some people, having a label is helpful epistemically by providing a category and framework for understanding their experience. These hermeneutical resources can support hermeneutical justice by enabling individuals who might otherwise be marginalized by their mental illness experience to understand their experience in meaningful ways (Wardrope 2015, 344–347). Having a label of mental illness can also be beneficial in a practical context by directing a person toward resources and tools that can help them deal with their experience, such as psychoeducation, therapies, and other forms of treatment, as well as access to disability accommodations. In addition, a label can help connect individuals with a community of others who share the same condition and have similar experiences, enabling both peer support and a community that can be involved with advocacy and activism. By connecting individuals with both resources and others in a shared community, labeling and medicalization can demarginalize otherwise marginalized groups of people (Reiheld 2010, 80–83). With a label of a specific mental disorder, or mental illness in general, a person can understand their experience in a mental health and medical context that makes sense of their thoughts, feelings, and actions and provides direction for how to manage or address aspects of their experience that they find problematic. These benefits can be empowering and liberating for some.

For other people, however, the label of having a mental illness can be devaluing and can be as harmful as the illness itself. Awareness of mental illness stigma can make a person want to reject having a mental illness label because they do not want to be associated with the negative attitudes and stereotyping that they perceive as accompanying a mental illness label. Not only can having a mental illness label make a person feel badly about themselves and their identity but it can also increase mental illness symptoms both by causing stress that exacerbates symptoms and through iatrogenic processes as people unconsciously adopt symptoms that they are believed to have.[1] Having a mental illness label can lead to a reluctance to seek help and adhere to treatment when a person does not want to identify with that label. Since the label of having a mental illness increases a person's vulnerability to stereotyping, prejudice, and discrimination, a person may have further reason to want to dissociate themselves from the label so they do not have to be subject to these (Gangi 2021, 388).

When a person with mental illness internalizes stigma, they may either self-label or recognize that others put a label on themselves in relation to the stigmatized trait. When other people stigmatize or have the potential to stigmatize an individual who has an undesirable trait, they put a label on the stigmatized individual as being different and as having the undesirable trait, which a person may internalize as to govern their thoughts, feelings, and behavior. With self-labeling, an individual labels themselves as being different from others and as having what they recognize to be a culturally undesirable trait. Self-labeling follows labeling by others. Individuals learn to self-label after observing others labeling them. A person does not come to recognize that they have a salient trait that marks them as different in kind from others without first observing that other people in their culture recognize that trait as different, and as different-making.

Annie Fox and Valerie Earnshaw describe a process in which individuals learn to self-label as having a mental illness (Fox and Earnshaw 2023, 2). First, they come to be aware of having symptoms – mental or behavioral processes – that aren't working the way they are "supposed" to, the way they "normally" do, or the way they did "before" (what Fox and Earnshaw identify as symptom awareness). Then they come to recognize that these symptoms are pathological in nature and may stem from an illness (what they identify as symptom appraisal). Upon becoming aware that these symptoms may be part of a mental illness, they learn to self-identify and

[1] Ian Hacking describes this phenomenon in terms of "the looping effect," where interactive kinds interact with their environment to become what they are (Hacking 1999, 103–119).

self-label as having a mental illness. When they become aware that their symptoms are symptoms of a specific mental illness, they may come to self-label as having the mental disorder that they could be diagnosed with, for example self-labeling as depressive or schizophrenic.

Labeling does not always impact how a person feels about their mental illness experience, but it easily can and often does.[2] For some, self-labeling can be empowering and provide resources, opportunities, and a connection with a community that are needed and otherwise would not be available. For others, however, self-labeling can invoke mental illness stigma that a person might want to distance themselves from.

Stereotyping

There are many stigmas and stereotypes associated with mental illness. Stereotypes of people with mental illness include the conception of them as being incompetent, the idea that they are unpredictable, the fear that they are dangerous, the view that they are unclean, and the attitude that they are worthless (Boysen et al. 2020, 346; González-Torres et al. 2007; Hipes 2016, 7; Mestdagh and Hansen 2014; 83; Riffel and Chen 2020, 1115). Other stereotypes include seeing people with mental illness as having an inherent weakness, believing that they lack willpower or are lazy (González-Torres et al. 2007, 17), viewing them as being immoral (Gangi 2021, 386), and regarding them as being insecure, unstable, and awkward to be around (Gangi 2021, 388). Some studies identify the three most dominant stereotypes about mental illness as involving incompetence, neediness and dependence, and dangerousness and violence (Corrigan and Kleinlein 2005, 16; Rüsch, Angermeyer, and Corrigan 2005, 530). Other dominant stereotypes are that people with mental illness are inherently defective or flawed, that they have a weak character in being unable to address their problems, and that their defectiveness is permanent with no hope for improvement or remission (Eminson 2004, 123; Gelder 2004).

Besides stereotypes of what people with mental illness are like, and what their condition of mental illness is like, there are also stereotypes and stigma related to treatment and help-seeking. For some people, there is stigma around taking medication and stigma around seeking mental health treatment, where those who seek treatment are regarded as being

[2] One study showed that how a mental disorder is labeled – as a disorder, an illness, a problem, and so on – played no role in a person's felt or anticipated stigma. People may internalize meanings of labels that can be placed upon them, but this does not always affect how they feel about their experience or what they expect from others who interact with them (Fox et al. 2021).

inherently weak for not being able to manage their problems on their own (Gangi 2021, 387–388). While stigma around help-seeking has shifted in recent years due to more positive views of mental health professionals and a recognition of the value of mental health treatment (Angermeyer et al. 2017; Onyeukwu and Donahue 2024, 432), norms of self-control, self-reliance, and personal responsibility continue to reinforce stigma against seeking mental health treatment for some people and in some communities (Courtney, Dial, and Yanos 2024; Lui, Sagar-Ouriaghli, and Brown 2024).[3]

Specific mental disorders have their own stigmas attached to them, often related to perceptions of competence and warmth.[4] In relation to the stereotype content model, people who have certain mental disorders are regarded as having different mixtures of warmth and competence. In one study, obsessive-compulsive disorder, generalized anxiety disorder, post-traumatic stress disorder, social phobia, depression, eating disorders, and panic disorder all made a person seem to have high warmth and low competence (Sönmez and Karaoğlu 2023). This led to emotional reactions of pity and sympathy toward these people, as well as paternalistic attitudes and treatment. Antisocial personality disorder, bipolar disorder, and borderline personality disorder were perceived as making a person have high competence but low warmth, creating emotional reactions of respect and envy in some contexts, and distrust, resentment, and even fear in other contexts (Sönmez and Karaoğlu 2023). Multiple personality disorder, schizophrenia, alcoholism, drug addiction, and anorexia all made a person appear to have low competence and low warmth, leading to the most derogatory emotional reactions that include disgust and contempt (Sönmez and Karaoğlu 2023). Illnesses with psychotic features, such as schizophrenia, score especially low on both warmth and competence (Görzig and Ryan 2022). In fact, one study found that people with all disabilities were rated with higher warmth and lower competence except schizophrenia, which was rated low in both (Canton, Hedley, and Spoor 2023). One study noted that mental disorders with more visible and severe symptoms were more disruptive of how a person behaves and interacts, making the person seem less warm; people with depression and bipolar

[3] For instance, see discussions of stigma in Asian American and Latinx communities assessed later in this chapter (Deng and Rose 2025; Helu-Brown, Barrio, and Hernandez 2023; Maeshima and Parent 2022; Misra et al. 2021; Morrow et al. 2020).

[4] Some theorists propose that negative stereotypes tend to conceptualize groups of people as low competence/high warmth, or high competence/low warmth, valuing one aspect and denigrating another (Cuddy, Fiske, and Glick 2002).

disorder, in this study, were viewed as less warm than people with anxiety (Follmer and Jones 2017).

One study on stereotypes of disability also measured courage, since some stereotypes of disability focus on how "brave" and "heroic" a person with a disability is perceived to be, suggesting that disability invokes courageousness (Canton, Hedley, and Spoor 2023). This study found that courage sometimes rated higher than warmth (as in anorexia) and sometimes lower than warmth (as in depression). In many disorders, warmth and courage were associated with emotions of pity, except for anorexia and depression, which garnered low levels of pity. Higher levels of competence, warmth, or courage all generated greater feelings of admiration (Canton, Hedley, and Spoor 2023, 493–494).

The gender associated with a mental disorder also plays a role in how the warmth and competence of a person who has the disorder are regarded (Boysen 2017). People who have disorders that are perceived to be masculine (such as antisocial behavior, substance abuse disorder, impulse control disorder, and paraphilia) are more likely to be regarded as lacking both warmth and competence. In contrast, people who have disorders perceived to be feminine (including eating disorders, body dysmorphic disorder, borderline personality disorder, histrionic personality disorder, and dependent personality disorder) are more likely to be seen as having more warmth. This association of warmth or lack of warmth with the perceived gender of a mental disorder occurs regardless of what gender an individual with the disorder actually expresses.

When a person internalizes stigma, they are aware of negative stereotypes and apply these to themselves, whether through endorsement of the stereotypes or through the way these shape thoughts, emotions, and behavior. Patrick W. Corrigan and Katherine Nieweglowski describe a four-stage model of self-stigma that includes being aware of a cultural stereotype, agreeing with the stereotype, and applying the stereotype to oneself, where this application leads to harm (Corrigan and Nieweglowski 2021, 417). The first three of these steps involve awareness of the stereotype, endorsement of the stereotype as true, and believing that the stereotype accurately describes oneself. If a person is unaware of the stereotype (an unawareness which would be hard to achieve in many cultures since negative stereotypes about mental illness are so prevalent in many cultures), they cannot evaluate it or apply it to themselves. If they do not agree that the stereotype is true, the stereotype could possibly lose its force and meaning so that a person would not think it matters and would not have a reason to apply it to oneself. If the person does not believe the stereotype accurately describes themselves, they might have no reason to be hurt by applying the stereotype to themselves because they would feel distant from it.

On the other hand, if a person does not agree that a stereotype is true, but they recognize that others might think or at least act as if it is true, they may feel that the stereotype will be applied to them regardless, and they may feel that they have to respond to this. If a person does not believe the stereotype accurately describes them, but they recognize that others might think that they conform to the stereotype, they may feel that they cannot escape the stereotype regardless of their own views about it and so they may feel constrained in having to respond to it. Having to respond to a stereotype because *others* apply it to oneself can involve just as much hurt and harm as believing that the stereotype is true and applies to oneself. Feeling that a stereotype is inescapable and feeling called to respond to it in some way or other is a way of internalizing stigma that others place upon oneself.

Stigma can be internalized in other ways besides the four-part model described by Corrigan and Nieweglowski (2021). The four-part model applies to self-stigma but not attendant stigma. People can believe a stereotype does apply to themselves without endorsing the stereotype as true; they do not have to agree with the stereotype in order to think it must describe themselves. They might not think it does describe them, or they may be uncertain about whether it describes them, but they may feel resigned to the fact that it *must* apply to them because others see it as applying to them. Moreover, people can disagree with a stereotype but be aware that others may believe it, or at least act and interact in ways that suggest they believe it, and so they see the stereotype as being applied to themselves even though they do not think it is true or accurate. When a person self-labels as having mental illness and/or is aware of the label that others put on them, and is also aware of the cultural stereotypes concerning what it is to have a mental illness, they are likely to feel resigned to the idea that those stereotypes must apply to them even if they do not actually believe the stereotypes and even if they do not believe they actually fit those stereotypes. Resignation does not require agreement with the stereotype or agreement that the stereotype accurately describes them.

Prejudice

Prejudice is a negative affective response to people that involves directing denigrative emotions such as fear, disgust, and contempt and pejorative reactive attitudes like shaming and blaming toward a person (Corrigan and Kleinlein 2005, 16–17). Prejudicial responses stem from a judgment that the people against whom one holds prejudice are inferior. People with mental illness commonly experience prejudice when they are feared as being dangerous or violent; when they are regarded as incompetent or overly needy and dependent, in which case they may be viewed with pity, contempt, and shame; or when they are viewed with disapproval due to intersecting stigmatized identities such as having a minoritized race, physical disability, or homelessness status. They are often blamed and shamed for bad actions they perform due to their illness, as well as for having undesirable thoughts, feelings, attitudes, behavior, grooming, and appearance associated with mental illness. Negative emotions and reactive attitudes commonly lead to shunning, avoidance, social exclusion, discrimination, microaggressions, and epistemic injustices.

The literature on stigma shows that people with mental illness are frequently subject to prejudice of different types in various settings. Prejudice in a work environment is common. Having knowledge of an employee's mental illness seemed to decrease an employer's confidence in the person as being able to do their job (Janda 2008, 424). According to one study, more than 40 percent of employers would not hire a person whom they knew to have a mental illness (cited in Janda 2008, 425). While bosses would be reluctant to hire or keep a person with mental illness as an employee, colleagues would also be unsupportive or even unwilling to work with individuals who have mental illness. Another study found that more than half of participants who did not have mental illness would not want to work with someone with depression, substance abuse, or schizophrenia (cited in Janda 2008, 425). Even when individuals with mental illness work in the mental health field as a support person with lived experience, they often experience stigma and discrimination by their colleagues (Byrne et al. 2019). For this reason, many people with mental illness try to conceal their illness from their employers (Schulze and Angermeyer 2003, 307).

Prejudice occurs in other settings as well. The fear that people with mental illness are dangerous and potentially violent makes many people unwilling to move to live near people with mental illness (Shearer et al. 2021). Students in all levels of education often perceive stigma from their peers, making the educational environment feel less safe for students with mental illness (Ahmad et al. 2020; Waqas et al. 2020). This also makes it difficult for students to seek mental health treatment, especially for themselves (Villatoro et al. 2022).

Healthcare professionals are also a source of significant stigma (Henderson et al. 2014). Many healthcare professionals fear people with certain mental illnesses as dangerous and see them as disorderly, incompetent, and needing perpetual help (Riffel and Chen 2020). They also feel frustrated by their powerlessness in being able to make a difference in the lives of people with mental illness (Riffel and Chen 2020, 1109–1110). Even though they work with people who have mental illness, mental healthcare providers are a significant source of mental illness stigma, especially in relation to disorders such as schizophrenia that are viewed as intractable, permanent, and incurable (Valery et al. 2023). There is also significant prejudice against people with mental illness in emergency room settings among physicians, residents, nurses, and staff (Smith et al. 2024).

In addition, people with mental illness experience stigma in the criminal justice system. Many lawyers have prejudices against people with mental illness based on various negative beliefs and stereotypes about them (Özcan and Burcu Ceylan 2022). Michael Perlin has documented many ways that the mental illness stigma he calls "sanism" manifests in the law (Perlin 2000; Perlin 1993; Perlin and Dorfman 1993). This includes pretextuality, where courts accept dishonest testimony in order to achieve a desired end; often, the false testimony harms people with mental illness (Perlin 2016). Issues such as the right to treatment, the right to refuse treatment, and involuntary civil commitment tend to be argued in the courts in ways that perpetuate various negative stereotypes and prejudices against people with mental illness.

Police officers frequently hold prejudice against people with mental illness, fearing such people as dangerous and subject to misinterpretation, and feeling a lack of empathy toward them (Tartaro et al. 2021, 580–581). Police who are CIT (Crisis Intervention Team)-trained are more stigmatizing than those who are not, perhaps because they see the extremes of mental illness behavior more in their training; they do, however, feel more prepared for dealing with people with mental illness (Tartaro et al. 2021). When people with mental illness feel like they are being treated fairly and with dignity through procedural justice, they are less likely to feel coerced and more likely to cooperate with police (Watson and Angell 2013; Watson et al. 2010). However, they are also more likely to cooperate when they perceive they are subject to greater stigma, perhaps because they feel motivated to act out of fear, a sense of survival, or demoralization (Watson and Angell 2013, 36). Black Americans are less likely to seek police help because of fear of being stigmatized and treated poorly because of their race (Altamirano and Weisman de Mamani 2024). However, when they have higher stigma and perceive a person with mental illness as more dangerous, this leads them to seek police help more (Altamirano and Weisman de Mamani 2024), perhaps because the dangerousness of the person with mental illness seems to outweigh the dangerousness of interacting with police.

People who have mental illness are subject to prejudice in nearly all facets of society and nearly all areas of their life. This has profound effects on their experience, on their expectations for others and for themselves, on how they see themselves, and on how they think, feel, behave, and interact with others. Prejudice impacts people's sense of self and identity and often leads to the various other forms of stigma.

When people with mental illness internalize stigma, they self-evaluate to determine whether they conform to negative stereotypes and thus whether they seem to be a fit candidate for stigma. If, in evaluating themselves, a person does believe negative stereotypes can apply to them, they then judge themselves as having a badness about them for possessing a trait that is seen as bad to have. This badness can make them feel fundamentally different and separate from others and lead to feelings of inferiority. The judgment that the self is marked by badness easily leads to negative reactive attitudes that a person directs toward themselves, including shame, guilt, and blame. This can also cause negative emotions associated with prejudice, including fear and disgust, to be directed toward themselves. In this way, a person applies prejudice toward themselves.

Just as others can direct negative emotions like fear and disgust, and negative reactive attitudes like blame, toward people with mental illness, a person with mental illness can direct these emotions and attitudes toward themselves. They may fear themselves because they fear having certain experiences, being seen in a certain way, and seeing themselves in this way. They may also fear themselves in believing that they will never get better, that they are inherently and permanently defective, that there is no hope that help or treatment could be effective, or that there is no hope for them. They may feel disgust toward themselves for having a mental illness and for having the thoughts, feelings, and behaviors associated with this. They may also feel disgust toward themselves when they appear to embody negative stereotypes that they would rather dissociate themselves from. People with mental illness often blame themselves for bad actions they performed due to their illness and for having undesirable thoughts and feelings, and they often feel incredible shame and self-loathing for having a mental illness and for having the thoughts, feelings, and behaviors associated with this.

Moral Distancing and Social Exclusion

People with mental illness are commonly subject to the moral distancing involved with separation and status loss, and the social exclusion that results from this moral distancing. Community members often morally distance people with mental illness by not treating them as credible, authoritative, and legitimate members of their community who can engage in the same social, moral, and epistemic practices that other community members engage in with each other. They are regularly avoided, shunned, and discriminated against, preventing them from having access to the same resources and opportunities, including social opportunities, that others have access to. This exclusion from participation in social, moral, and epistemic practices inhibits the ability of people with mental illness to feel a sense of belonging that is so crucial to human experience and identity. It also reinforces the view that people with mental illness have less moral status and are less of an autonomous agent, less of a person, and less of a human being.

Social exclusion causes many harms. Being left out of social interactions is associated with older age; less education; greater hospital admissions; more severe symptoms, which disrupt spheres of activity such as work, social interactions, and social relationships; and experienced and anticipated stigma (Mezey et al. 2022). In addition, social exclusion exacerbates mental illness symptoms and inhibits recovery. In a vicious circle, mental illness symptoms heighten social exclusion, and social exclusion worsens mental illness symptoms (Reinhard et al. 2020). As people feel uncomfortable or even fearful engaging with someone who has active mental illness symptoms, they have a tendency to avoid social interactions with a person who is actively symptomatic, making the person more socially isolated. Increased social isolation and exclusion create anxiety, depression, and stress for the person with mental illness and exacerbate whatever mental illness symptoms they have. This vicious cycle is hard to get out of and requires intentional action on the part of people who have to overcome their discomfort interacting with those with mental illness as well as people with mental illness who have to make attempts to reach out despite their social exclusion so they can engage with others more fully.

People with mental illness have fewer close relationships, including intimate relationships, in part due to the difficulties with interaction that mental illness symptoms create, but also in part due to mental illness stigma (Adu et al. 2023, 227; Mestdagh and Hansen 2014, 83; Thornicroft et al. 2022, 1447). Losing friendships in the throes of illness is common (Thornicroft et al. 2022, 1447). They also have fewer social contacts in general due to social and moral distancing (Boysen et al. 2020, 346). Even within the relationships that they have, the quality of their relationships sometimes suffers due to their illness, and they can experience stigma even within these relationships.

People with mental illness are not only socially excluded and outcasted from society but they are also subject to moral distancing and exclusion within their families. Social exclusion and estrangement from family is not uncommon (Thornicroft et al. 2022, 1447). Even when family members do not exclude a person, they propagate stigma through several means, including how they talk *to* the individual with mental illness, how they talk *about* the individual with mental illness with other people, and when they dissociate themselves the person or become estranged from them (Adu et al. 2023, 224). Psychosocial effects of family stigma include feeling shame and embarrassment, both for the person who has the mental illness and for the family members; having to live with perceived danger and threat from the person with mental illness; various forms of family conflict; and neglect by one's spouse, parents, or other close family members (Adu et al. 2023, 227–228).

As their social contacts decrease, people with mental illness often find that most of their social interactions occur with other people who have mental illness (Mestdagh and Hansen 2014, 84). This is because they are more likely to interact with others who are experiencing similar life events as them. For example, they may socialize with other patients receiving mental healthcare treatment, as when patients meet and interact with other patients, such as through a support group or behavioral health program. Or, if they are homeless, they are more likely to socialize with other homeless individuals who may be mentally ill themselves.

As social contacts in other areas of their life drop away, the social interactions and relationships that are left tend to be with others who are mentally ill like themselves. This leads to a homogenous range of social contacts, comprising of individuals like oneself, and excluding many individuals who are different from oneself as would occur in a heterogenous range of social contacts. Having only a homogenous range of social contacts is advantageous in some ways and detrimental in other ways. Interacting with people like oneself can provide important sources of support of people who understand what one experiences because they have gone through something similar, and thus who can empathize well and sometimes show compassion better.

However, it is detrimental in other ways. Interacting only with people like oneself limits what resources a person has access to that they can draw upon in support. When a person interacts with others who are unlike themselves, they have access to a broader range of support and resources than what they can access on their own. However, when they interact only with others like themselves, they miss out on this broader range of access and so have fewer resources they can draw upon. This limits the kinds of support that are available to them and also constricts social mobility. When individuals have social contacts across the social classes, they are able to move across social classes themselves, enhancing social mobility. When individuals have social contacts only in one social class, they tend to stay in that social class themselves and lose out on not only the resources available in other social classes but also the different kinds of knowledge that are particular to different social classes. Homogenous social contacts thus lead to diminished support and resources, lack of social mobility, and a constricted range of knowledge.

It is not only the shunning, avoidance, and discrimination of others that create social exclusion and social isolation. Anticipated stigma also easily leads to concealment, social withdrawal, and consequently isolation. Individuals with mental illness often fear being discriminated against, so they may try to hide their diagnosis, symptoms, and medication use from others, hoping that others will be unaware of their label, their problems, their distress, and their treatment (Mestdagh and Hansen 2014, 83). If they can succeed in hiding their illness, label, and treatment from others, then, they hope, others will treat them normally and refrain from stigmatizing them. When concealment doesn't work, people who anticipate stigma from others may socially withdraw from others in order to avoid others who may be stigmatizing and thus to avoid stigma itself. Social withdrawal can lead to the same isolation that moral distancing and social exclusion does.

Efforts to include people with mental illness in social interactions and in the broader community are thus extremely important. Social inclusion corresponds with less loneliness and better quality of life (Mezey et al. 2022, 423). Some of the factors that contribute to social inclusion include participation in social activities, having various kinds of social supports, having adequate and safe housing, being nested within a neighborhood, being able to be involved with the community, having some advanced education, having a job, having adequate physical health and well-being, and using mental healthcare services (Filia et al. 2019). Eliminating experienced stigma and discrimination is also vital. In addition, reducing anticipated stigma by creating an environment where people with mental illness do not fear that they will be stereotyped, stigmatized against, and discriminated against by others is also essential.

When people with mental illness internalize stigma, they often separate themselves from others and from society in general. They engage in social withdrawal, feeling as if they do not belong with others and so consequently avoid others; this causes social isolation. Sometimes this avoidance occurs because a person does not feel they deserve to belong, where they isolate themselves on account of feeling fundamentally different from others. Sometimes it occurs preemptively so a person feels like they have control over their situation: they try to avoid others before others can avoid them, so they can have power over what happens to them. Willful social isolation is a common response to the internalization of stigma, especially mental illness stigma.

In addition to social withdrawal, people with mental illness often refrain from participating in various activities that they would normally participate in. Refraining from participating in activities can result from a person feeling that they do not deserve to participate or deserve to receive the benefits of participation, or from feeling like their participation would harm others, or from feeling like an outcast who lacks the belonging that seems necessary for participating in the activity. In preventing themselves from participating in activities they would normally participate in, a person engages in a sort of self-discrimination, excluding themselves from activities on account of their stigmatized trait.

Status Loss and Dehumanization

People with mental illness often experience status loss. They are viewed as having less agency, less autonomy, less rationality, less control over themselves and their behavior, less control over their thoughts and feelings, less control over their self-presentation and how they come across to others, less morality, less personal identity, less as persons, and less than human. This status loss typically results in dehumanizing people with mental illness.

Dehumanization consists of viewing individuals as lacking characteristically human traits such as morality, rationality, self-control, intentional choice, agency, and culture, and comparing individuals to monsters, animals, machines, or tricksters rather than seeing them as full humans (Boysen et al. 2020, 347). When they are compared to creatures or objects, individuals are seen as lacking warmth, depth, capability, agency, autonomy, an identity of their own, rationality, morality, personhood, and humanness. Studies indicate that people generally regard those with mental illness as significantly less human than other marginalized groups because they lack many of these traits (Boysen et al. 2020; Burguet and Girard 2024). Viewing someone as subhuman in this way involves seeing them as lesser compared to others.

One study showed that people actively dehumanize those with mental illness, viewing them as "significantly less human" compared to other dehumanized groups such as stigmatized racial and ethnic groups (Boysen et al. 2020, 346). In this study, stereotypes of unpredictability, incompetence, and dangerousness of people with mental illness were invoked. Study participants exhibited prejudice through emotional reactions of fear and anger, and automatic negative associations with mental illness. People with mental illness were dehumanized in the sense that they were viewed as lacking characteristically human traits such as warmth, depth, agency, competence, morality, rationality, and culture, making them seem more like animals or machines. Contrary to some models of dehumanization, dehumanization can occur regardless of how "warm" a person seems, especially when their competence is in question (Boysen, Chicosky, and Delmore 2023). This seems to be the case with many mental illnesses.

When stigma is internalized, people with mental illness conceptualize themselves as losing status as they learn to judge themselves as being inferior to others on account of their mental illness. This self-judgment arises from social awareness of others' judgments. People only learn to see a socially salient trait as the basis for an evaluation of being inferior when they observe others evaluating those with the trait as inferior. Without the cultural and social context of seeing how other people view the trait as inferior-making, it might never occur to a person that the trait should make them inferior to others. After all, for many traits, there is nothing inherent to the trait that makes those who possess the trait appear inferior; it is only people's judgment that makes the trait seem salient in a way that marks the person as different and inferior.

The presence of negative stereotypes within a culture automatically marks people who appear to have the stereotyped trait as inferior and makes them seem lesser compared to others. Self-labeling, like labeling by others, always automatically invokes these negative stereotypes, reinforcing the devaluation of the person who has the relevant social trait and thus marking the person as lesser compared to others, in the person's own eyes as well as in the eyes of others. When a person with mental illness internalizes stigma, they see themselves – or see themselves through the eyes of others – as less sane, less rational, less autonomous, less "put together," with diminished functioning and agency, less moral, less of a person, and less of a human being. This awareness of being lesser compared to others easily leads a person to feeling a lack of belonging with others and commonly results in social withdrawal and self-isolation. It also leads to judging oneself as bad and reacting to oneself with shame, guilt, and blame. Thus, this feeling of inferiority reinforces the prejudice and social withdrawal that a person directs toward themselves when they internalize stigma.

Microaggressions

People with mental illness commonly experience microaggressions, where those who interact with them have implicit biases against mental illness that they are not aware of and consequently speak and act in ways that harm or even dehumanize people with mental illness. One study identified three themes of microaggressions: those that assumed that people with mental illness are inferior; those that invalidated the experience, perspective, and identity of people with mental illness; and those that perpetuated fear of mental illness (Gonzalez et al. 2015). Another study delineated four themes of microaggressions (Peters et al. 2017). Some perpetuated stereotypes of people with mental illness as inferior, as seeking attention, as being cold or immoral, as being responsible for causing their own plight, and as using mental illness as an excuse. Mirroring the former study, some microaggressions invalidated the experience and perspective of people with mental illness by being suspicious of its existence or its severity. Other microaggressions viewed individuals with mental illness entirely through the lens of their illness, basically defining their identity by their illness, while still others involved the misuse of terms like "crazy" and "psychotic" in contexts where these terms were not appropriate.

In the context of mental healthcare services, one study found three themes of microaggressions (Defourny et al. 2024). Those that dehumanized people with mental illness failed to treat individuals as whole persons but instead as problems needing fixing, and they tended to blame and shame individuals. Some microaggressions committed epistemic injustices of disregarding the perspectives of people with mental illness and failing to share information and decision-making with them, likely with the belief that individuals with mental illness were not capable of being equal participants in the epistemic practices involved with mental healthcare. Other microaggressions portrayed the situation of people with mental illness as hopeless by emphasizing the chronicity of mental illness, failing to set reasonable goals for clients, and failing to engage them in meaningful or even everyday activities in hospital settings.

People who are closest to those with mental illness, such as family, friends, and mental healthcare providers, both had the most stigma and committed the most microaggressions and had the least stigma and committed the least microaggressions compared to people who were not close to those with mental illness; this suggests that there is a u-shaped, curvilinear relationship between familiarity and stigma (Gibson, Lindsay, and Szeto 2025). Interestingly, individuals with mental illness perceived more microaggressions aimed at them by those closest to them, such as family members, friends, and mental healthcare providers, while those closest to them believed they committed fewer microaggressions when they had more contact with people with mental illness (Gibson, Lindsay, and Szeto 2025, 3; Peters et al. 2017). This suggests that many people are unaware that they commit microaggressions and believe they have supportive attitudes toward people with mental illness, obscuring implicit biases they may hold.

Discrimination

Discrimination can be understood as differential treatment in which a person is treated in worse and unequal ways compared to others. Discrimination is a behavioral aspect of stigma, having to do with how people act toward those who are stigmatized against. People with mental disorders commonly experience discrimination in many different areas of their lives.

In one study, almost 90 percent of participants experienced discrimination in at least one area, including close partner relationships, parenting, finding and keeping a job, and education (Farrelly et al. 2014, 7). In each of these areas, more than half of all participants experienced discrimination in that area in the past year. In another study, about one-third to one-half of participants experienced discrimination in the contexts of family, making or keeping friendships, intimate relationships, and finding and keeping a job (Thornicroft 2009). A third study noted that discrimination is common in areas including healthcare, employment, criminal justice, media, and public participation (Voldby 2022, 5). Whether it is 33 percent or 90 percent of groups of individuals with mental illness experiencing discrimination, this is a significant amount of discrimination that is occurring.

Some of the areas where individuals experience discrimination include major life domains such as attaining and keeping a job, finding a place to live, receiving education and training, and accessing healthcare. Employment discrimination is common. Twice as many people with mental illness report employment discrimination as people with physical disabilities (cited in Stuart 2006, 523). One-third of people with mental illness surveyed reported being rejected from being hired or even having a job rescinded when their mental illness was made known (cited in Stuart 2006, 523). Surveys of employers show that half of employers would be reluctant to hire someone with a known mental disorder and would even be willing to fire someone once their mental disorder became known (cited in Corrigan and Kleinlein 2005, 18–20; Stuart 2006; Thornicroft 2006). One survey indicated that people with mental illness receive fewer invitations to be interviewed when their illness was known, based on a belief that people with mental illness are fragile and unreliable (Østerud 2023). The *Lancet* Commission on Ending Stigma and Discrimination in Mental Health found that people with mental illness have substantial difficulties finding employment; that they do not have equal opportunities when employed; that they are assigned less work and limited responsibilities in their jobs; that they are often denied promotions; that they experience bullying at work; and that they are often not given time off to attend needed healthcare appointments (Thornicroft et al. 2022, 1447).

Some of the negative stereotypes about mental illness that employers endorse include that they are incompetent, dangerous, unpredictable, or do not have the capacity to work. Other negative beliefs include the idea that mental illness is not a legitimate illness, that people with mental illness will shirk their responsibilities so that others will have to pick up the slack, that work is bad for the mental health of a person with mental illness because of the stress it causes, and that employing people with mental illness is an act of charity rather than a legitimate business decision (Krupa et al. 2009). Still other beliefs are that people with mental illness are not readily employable, that they are not capable of competitive employment, and that they do not want to work (Rebeiro et al. 2012, 13). These stereotypes and beliefs all make employers reluctant to hire people with mental illness as workers. When people with mental disorders are hired, it is much more likely to be for low-skilled jobs; people with mental illness have very low rates of being employed in high-skilled jobs (cited in Stuart 2006, 523). In a vicious cycle, people with mental illness who are unemployed experience increased stigma and discrimination due to the compounding nature of their employment status and their mental health status (Oudejans, Spits, and van Weeghel 2021).

People with mental illness also face housing discrimination. Surveys of landlords report that many landlords are less willing to rent to people with known mental illnesses compared to the general population (Corrigan and Kleinlein 2005, 18–20; Thornicroft et al. 2022, 1447). A HUD (US Department of Housing and Urban Development) report found that people with mental disabilities are less likely to have their rental inquiries be adequately responded to, to be told about available units, to be asked to contact the housing authority, and to tour an obtainable unit; and more likely to be directed to a different unit than the one advertised and to receive differential treatment based on the type of disability they have (US Department of Housing and Urban Development (HUD) 2017). People with mental illness who experience high stigma and discrimination are much less likely to have housing stability and more likely to be homeless (Mejia-Lancheros et al. 2021).

Another place where people with mental illness face discrimination is in education. Both Stanford University and Yale University were sued for having policies that automatically withdrew students from school who did not withdraw voluntarily after their mental health problems were known (Wan 2022).[5] Universities that have policies like this tend to do so for liability reasons, to prevent them from being sued or seen as criminally negligent if a student causes harm to themselves or others in a mental health crisis. However, involuntary withdrawal is a discriminatory and punitive approach to students seeking mental health help. Students can also be discriminated against in an education setting when educators refuse to provide the accommodations students need in order to be successful in school, or when universities do not admit prospective students because of known mental health problems. When students are viewed as a liability rather than as a human being with potential for growth and development, their mental health conditions are often used against them.

Stigma also leads people with mental illness to withdraw from education and work on their own (Thornicroft et al. 2022, 1447). Sometimes people

[5] In Yale's case, the involuntary withdrawal meant that students wanting to return to Yale after getting psychiatric help had to reapply for admission. After being sued, Yale changed its policy to allow students to take a medical leave of absence where they did not have to reapply but could instead be readmitted following a letter explaining why they left, what treatment they received, and why they think they are ready to resume coursework. This seemed fairer to students, and also allowed students to keep their student health insurance so they could receive the psychiatric treatment that they needed to get better.

leave jobs and educational programs because they feel they cannot complete their studies or fulfill their work responsibilities due to their mental illness symptoms, but anticipated discrimination exacerbates this. When people fear they will be rejected and that their needs will not be accommodated in workplace or educational settings, they are more likely to feel resigned, not having the energy to face or fight anticipated discrimination. They leave on their own due in part to anticipated discrimination before they can be actively discriminated against and feel forced to leave.

Stigma and discrimination also occur in healthcare. While mental health professionals generally have less stigma toward people with mental illness than the general public does, they do sometimes still stigmatize the clients they work with, especially those with mental disorders that are more severe, seem more incurable, and are perceived as more dangerous and more blameworthy (Bizumic, Gunningham, and Christensen 2022). As a result, mental health professionals may choose jobs where they don't have to work with such clients, and they may avoid these clients when they do encounter them in their work. In healthcare settings, people with mental illness are often not believed and not taken seriously, and the seriousness of their symptoms is often undermined (González-Torres et al. 2007, 18), leading to epistemic injustices of not being taken seriously as a knower and as a provider of credible testimony. Healthcare providers also stigmatize and discriminate against those who are homeless, especially when dealing with patients who get discharged from hospitals into homeless conditions (Canham et al. 2022). Stigma by healthcare providers, in the context of both mental and physical health, leads to service avoidance in order to avoid being stigmatized against, and consequently worsened physical and mental health (Reilly, Ho, and Williamson 2022, 2139).

People with mental illness are commonly shunned and avoided. People also experience what González-Torres et al. call "daily social discrimination," where they are relegated to the end of the line, where they are not respected at work and in other spaces, and where they are treated as weird or are pitied (González-Torres et al. 2007, 18). In many contexts, they are discredited, disbelieved, dismissed, ignored, and undermined. In one study, family members reported that they witnessed discrimination of the family member with mental illness in several areas, including prejudice by others, employment discrimination, discrimination by friends, seeing the person isolated, and wanting to overprotect them (González-Torres et al. 2007, 19). Discrimination can occur in any context where people can favor or disfavor some over others.

Anticipated discrimination is even more common than experienced discrimination. In one study, over 90 percent of participants anticipated and feared being discriminated by others (Farrelly 2014, 8). In another study, half to three quarters of participants anticipated discrimination in seeking close relationships and applying for work, job training, or education (Thornicroft 2009). In this study, over a third of participants anticipated discrimination even when they had not experienced it, suggesting that anticipated discrimination is persistent even in the absence of actual discrimination. Why this is so is explored in Chapter 3, where I look at how and why stigma persists even when individuals do not actively engage in it.

Discrimination causes both distributive injustice and structural injustice. When people are denied access to resources and opportunities such as housing, jobs, education, and adequate healthcare, they receive an unequal and unfair distribution of these resources and opportunities. This can prevent them from getting their basic needs met. This prevents them from achieving standards of justice such as preference satisfaction; living a life of equal desirability; living a life based on their own conception of the good; having the opportunity to hold social roles, positions, and offices; having agency and autonomy; having their human rights fulfilled; and living a worthwhile or at least minimally decent life.

Discrimination also contributes to structural injustice. Unjust social structures prevent people from accessing resources and opportunities such as healthcare, education, employment, and housing. Some of these unjust social structures include a housing market that requires considerable capital to participate in; an educational domain that requires financial means to access; a healthcare system that often requires a job to gain access to the system; and an employment market that requires education, training, and experience that many people do not have financial or practical access to. Unjust social structures also include systems and operations that make it difficult for individual and institutional agents to include those with mental illness in their social, epistemic, and moral practices, and that make employment, housing, and other forms of discrimination easy. When social structures are arranged in a way that encourages people and institutions to discriminate and exclude certain groups of people, it is all too easy to do so, and it can be very difficult to resist actions that discriminate and exclude, creating obstacles that are hard to overcome. Resisting discrimination so as to create conditions of structural and distributive justice often requires considerable conscious intention, energy, and effort, and only those who are dedicated to this end tend to put in the labor required to work toward it.

Epistemic Injustice

People with mental illness are subject to many types of injustice, including epistemic injustice, distributive injustice, and structural injustice. Distributive and structural injustices tend to result from discrimination, but epistemic injustice is a distinctive form of injustice that results from all aspects of stigma, not just discrimination. Epistemic injustice is connected to and results from labeling, stereotyping, prejudice, moral distancing, social exclusion, loss of status, dehumanization, microaggressions, and discrimination.

Havi Carel and Ian James Kidd describe many epistemic injustices that people with illness experience. Pathocentric epistemic injustices are epistemic challenges, needs, threats, and vulnerabilities (or "predicaments") that ill people experience (in other words, predicaments that they experience as a knower) in the context of their illness (Kidd and Carel 2021). Correspondingly, we can understand the epistemic injustices that people with specifically *mental* illness experience as *psycho-pathocentric* epistemic injustices. People with mental illness experience psycho-pathocentric epistemic injustices that stem from identity-prejudicial credibility deficits due to their social identity as a person with mental illness. Both pathocentric and psycho-pathocentric epistemic injustices can be transactional, institutional, and structural.

Transactional and institutional psycho-pathocentric epistemic injustices occur when individual and institutional agents interact with people with mental illness in ways that discount their testimony, discredit them as epistemic agents, deny them the opportunity to participate in various epistemic practices, and preclude them from having the hermeneutical resources needed to make sense of their experience. Some of the psycho-pathocentric testimonial injustices that people with mental illness commonly experience include epistemic objectification, where people with mental illness may be valued as sources of information, yet excluded from epistemic participation such as the ability to interpret their information (Dotson 2008); participatory prejudice, where people with mental illness are seen as lacking the capacities needed for having a sense of relevance and thus viewed as incapable of providing relevant and meaningful epistemic contributions (Carel and Kidd 2017, 339–340; Hookway 2010, 157–158; Kidd and Carel 2017, 180–181); and testimonial prejudice, where people with mental illness are viewed as unreliable reporters because it is assumed that they are globally incompetent and thus untrustworthy as knowers. These testimonial injustices often stem from pathophobia, which are the negative attitudes toward and treatment of people with illness based on their illness (Kidd and Carel 2021, 74). Pathophobia tends to involve five types of attitudes and behavior, including aversion, banality, callousness, insensitivity, and untruthfulness, all of which influence how individual and institutional agents interact with ill people, including people with mental illness (Kidd 2019; Kidd and Carel 2021).

People with mental illness can be epistemically discredited in other ways as well. They often have a style of expression that does not fit norms of expression of being calm, rational, and self-controlled (Carel and Kidd 2017, 342; Kidd and Carel 2017, 184–185). This can lead them to not being respected and not being taken seriously. For example, people with mental illness may be highly emotional, whether this is due to their illness or to being in a situation that is distressing, such as being hospitalized, which can make a person anxious, angry, or sad (Kurs and Grinshpoon 2018, 340). Moreover, a person with mental illness may have disorganized speech, which can lead others to assume that the content of their speech is meaningless and consequently gets dismissed as confused, irrational, and lacking sanity (Sanati and Kyratsous 2015, 482). When a person is believed to be delusional, healthcare providers often interpret most or all of the person's speech and action as a product of their delusion (Sanati and Kyratsous 2015, 481). Thus, the content of their speech tends to get dismissed as lacking relevance and meaning (Kurs and Grinshpoon 2018, 340; Sanati and Kyratsous 2015, 482). This can be quite damaging; for example, reports of abuse, neglect, or other harms may be misconstrued as delusions and therefore not adequately addressed (Pilling 2021, 145–150). Furthermore, the behaviors and emotions of a person with mental illness are commonly construed as symptoms of their mental illness, so that when they share testimony of experiencing pain, distress, or a physical ailment, this testimony is disregarded, and their experience is misinterpreted as mental illness symptoms and not adequately addressed (Carel and Kidd 2014, 534; Kurs and Grinshpoon 2018, 337).

Another place where people with mental illness are disregarded as credible knowers is when institutions and structures such as hospitals, behavioral health clinics, disability systems, and housing systems have policies and practices that deny their clients any voice over what happens to them within these settings. When they do not involve clients in decision-making and do not have mechanisms to elicit feedback from clients, they preclude their clients from having a say over what happens to them and thus deny them a significant source of agency. In addition, people with mental illness can experience testimonial injustice when healthcare providers seek limited information from them, such as by asking yes-or-no questions, without trying to understand the nuance and complexity of their condition (Carel and Kidd 2014, 534).

In addition to transactional and institutional psycho-pathocentric epistemic injustices, people with mental illness may also experience structural psycho-pathocentric epistemic injustices. One place where this occurs is in the way the medical and mental health systems uphold a naturalistic approach to illness that conceives of illness as observable facts that can be measured, recorded, and assessed. This is opposed to viewing illness in phenomenological terms as the lived experience of people and what it is like for them to have their illness (Carel 2019, 9–15; Kidd and Carel 2018). When people with mental illness are given a naturalistic approach to understanding their condition, as often happens in medical and mental healthcare settings, they are often prevented from having access to phenomenological tools that could better help them understand what they are going through. This leads to the hermeneutical injustice of being denied adequate hermeneutical resources that allow them to make sense of their experience.

People with mental illness also experience structural psycho-pathocentric hermeneutical injustice through what Şerife Tekin calls a "DSM culture," in which mental illness experience is understood primarily in terms of symptoms, or DSM (*Diagnostic and Statistical Manual*) criteria (Tekin 2014; Tekin 2011). This is a reductive way of understanding the mental illness encounter that ignores social and relational aspects of a person's experience (Lakeman 2010, 152l; Tekin 2014; Tekin 2011). When individuals understand their own experience in this way, this leads to impoverished self-insight and hyponarrativity, or a very thin narrative that ignores the richness, depth, nuance, and complexity of experience (Tekin 2014; Tekin 2011). Women may have a particular tendency to view their experience in this reductive way because of the way they are socialized to interpret their agency and what happens to them (Hoffman and Hansen 2017, 294). DSM culture is sustained through structural practices of healthcare systems, the psychiatric and psychological professions, pharmaceutical companies, and other institutions and individual agents that reproduce DSM understandings of mental illness, influencing media and culture to the extent that individuals within the culture absorb and perpetuate these understandings themselves.

When people with mental illness internalize stigma, they can impose self-directed epistemic injustice onto themselves through self-silencing, which is a form of epistemic exclusion. Kristie Dotson identifies two types of self-silencing, testimonial smothering and self-censoring (Dotson 2011). Testimonial smothering occurs when a person who wants to offer testimony believes that their potential audience is unwilling or unable to respond to their testimony appropriately, so they withhold their testimony. People with mental illness may believe that healthcare providers, colleagues, family or friends, or people in the general public will not respond appropriately to what they have to say, so they may suppress what they want to share. This especially occurs when they believe they will be misunderstood if they speak about their experience (Steslow 2010, 32). Self-censoring is another type of self-silencing, which occurs when a person loses confidence in themselves as a speaker and believes they have nothing worthwhile or relevant to share (Kidd and Carel 2017, 177). When individuals are treated by others as having nothing relevant or meaningful to contribute, they may internalize this perception and believe it of themselves, too, leading them to withhold whatever they might otherwise want to share.

People with mental illness commonly experience all components of stigma, including labeling, stereotyping, prejudice, moral distancing and social exclusion, loss of status and dehumanization, microaggressions, discrimination and the distributive and structural injustices that result, and epistemic injustice. These cognitive, affective, and behavioral processes of stigma occur both in public stigma, which is directed at a person by others, and in internalized stigma, which is projected by a person onto themselves. These aspects of stigma cause considerable harm to people with mental illness in the numerous ways delineated here.

Now that we have sketched out what mental illness stigma looks like, let us examine some of the factors that influence how this stigma occurs and how it is imposed on people with mental illness.

Factors that Impact Mental Illness Stigma

Many factors affect the kind, degree, and scope of stigma associated with mental illness. These factors include beliefs, political values, cultural values, socioeconomic status, education, and gender. There is extensive literature studying the effects of these factors on mental illness stigma, which I summarize here.

How Beliefs, Political Values, and Cultural Values Contribute to Stigma

Two kinds of beliefs in particular increase the desire for social distance from people with mental illness: the belief that people with mental illness are needy and dependent and cannot take care of themselves, and the belief that people with mental illness are incompetent, innocent, and childlike (Corrigan et al. 2001). These beliefs lead a person to think that people with mental illness are irresponsible or incapable of taking responsibility for themselves, which violates neoliberal values of self-sufficiency and individual responsibility which are prominent in American culture. This can cause those who hold these beliefs to be angry and resentful of people with mental illness, increasing the desire for social distance and increasing stigma and prejudice (Corrigan et al. 2001, 223).

Neoliberal values of self-sufficiency and individual responsibility are prioritized among both those who are politically conservative and politically liberal. Both conservative and liberal values can contribute to stigma against people with mental illness. Those who hold conservative values are more likely to see mental illness as a threat and to believe that the world is a just place where people get what they deserve, so that people with mental illness must have made bad choices or done something bad to "deserve" their illness. Those who hold conservative values also tend to endorse the Protestant ethic ideal of working hard, with the belief that working hard enables one to achieve success (Schomerus and Angermeyer 2021, 3).

Perceived blameworthiness also increases stigma and desire for social distance. When people with mental illness are regarded as blameworthy for their illness, the general public desires greater social distance from them compared to when their illness is viewed as occurring through no fault of their own (Markowitz and Syverson 2021, 921). Because people who are politically more conservative view people with mental illness and homeless individuals as blameworthy for their conditions, they tend to desire more social distance from these individuals compared to people who are more politically liberal (Markowitz and Syverson 2021, 927).

Sometimes holding liberal values decreases stigma by increasing tolerance, benevolence, and open-mindedness toward people with mental illness. In their study on the role that cultural values play in mediating mental illness stigma, Rieckhof et al. found that people who espoused liberal values of tolerance of others and openness to change had the most positive views of people with mental illness, in part because they regarded their experience of dealing with those with mental illness as enriching (Rieckhof et al. 2021).[6] This made them more open to interacting with these individuals and holding less stigma toward them. In contrast, people

[6] Tolerance of others and benevolence constitute what Schwartz (2012) calls "values of self-transcendence"; self-transcendence and openness to change are liberal values. In their study, Bojanowska and Kaczmarek (2022) found that self-transcendence and openness to change increase a person's well-being when activated through behavior that employs these values. Security,

who held authoritarian or conservative values of security had the most negative views of people with mental illness, associating them with dangerousness and violence. This activated their belief that these individuals were threatening and consequently increased their desire for social distance from these individuals.

Nevertheless, people who hold liberal values can still have beliefs that contribute to stigma. People with mental illness can seem to violate the values of self-actualization and achievement when they are unable to take action to change their circumstances and overcome their illness. People with mental illness can seem to violate the value of social/economic/location mobility when they are too dependent on families or social systems to move. And they can violate values of creativity, self-reliance, and independence when they appear incompetent, needy, dependent on others, and unable to change their circumstances. People who hold liberal values often have assumptions about normalcy in which normalcy involves being flexible, productive, and positively minded; mental illness seems to violate these norms when people with mental illness are unable to be flexible due to mental illness symptoms, unable to be productive when their competence is challenged, and have a negative mindset due to distress and despair caused by their illness and their social circumstances (Schomerus and Angermeyer 2021, 3–4).

Many cultural values also contribute to stigma, especially among people of color and other marginalized groups. Black Americans, Latinx, and Asian Americans have higher rates of stigma than the general US public, endorsing a variety of stereotypes about mental illness, including stereotypes of incompetence, dangerousness, and inferiority (Subica and Link 2024). They desire social distance from people with mental illness at a greater rate, and they have less confidence that mental health treatment will improve the condition of people with mental illness. Partly as a result of this (and also as a result of anticipated stigma based on their race), they seek out treatment for mental health problems at a much lower rate than White Americans (Subica and Link 2024).

Some studies show that mental illness stigma is higher in minority racial and ethnic groups than in White groups because of the presence of cultural beliefs and expectations that shape how an individual is perceived and how they see themselves (Misra et al. 2021; Wong et al. 2018). For instance,

conformity, and tradition constitute what Schwartz calls "values of conservation" or conservative values. Bojanowska and Kaczmarek observed that conservative values increase self-esteem by providing a robust grounding for thinking well of oneself within one's community, but behavior that employs these values does not increase well-being in part because it can create conflict with others (the "out-group").

cultural values and expectations about masculinity affect how people internalize stigma and how they apply stigma to themselves as not living up to the values and expectations that are part of their culture. As one example, Asian men may draw from their culture the value of working hard and not wanting to be a burden to others, idealizing strength, stoicism, self-sufficiency, and being a provider. As a consequence, they may regard receiving help for conditions like mental illness to be an indicator of weakness (Morrow et al. 2020). In addition, they may also value emotional restraint, avoiding shame, saving face, and avoiding self-disclosure and emotional expression, all of which also inhibit seeking professional help for mental health difficulties (Maeshima and Parent 2022).

As another example, in some Latinx cultures, people with mental illness are commonly viewed to be dangerous and out of control, and there is a belief that mental illness is permanent and incurable, leading to stigma, social distancing, prejudice, and discrimination (Helu-Brown, Barrio, and Hernandez 2023, 345). Some aspects of Latinx culture contribute to stigma, including the way communities tend to have tight bonds (where mental illness could be seen as contagious among members), acceptance of power differentials in society (and acceptance that people with mental illness have less power and privilege than others), lack of ambiguity and strict social norms (where mental illness is seen as deviance from these norms), and an emphasis on masculine norms such as strength and emotional reserve (Louden et al. 2023). Latinx people with mental illness experience many harms as a result of mental illness stigma within their cultural context. For example, social distancing and rejection, along with high rates of unemployment among the mentally ill, are primary obstacles to Latinx people with schizophrenia being able to develop and maintain romantic and other social relationships (Helu-Brown, Barrio, and Hernandez 2023). Latinx people experience relatively high rates of mental illness but have lower rates of treatment, due largely to fears of stigma related to both their mental illness and their race (Fripp and Carlson 2017).

While political and cultural beliefs and values can contribute to stigma, they can also protect against stigma, depending on what they are. For example, a belief that people are interconnected, which is a belief that Buddhists hold, leads to a greater sense of obligation toward others, greater empathy, and greater desire for understanding of others. This contributes to people being better able to aid those who are vulnerable, including people with mental illness (Yu, Mak, and Lai 2025). When political and cultural beliefs support empathy, understanding, and acceptance of others, this reduces and protects against stigma.

How Socioeconomic Status and Education Contribute to Stigma

People with higher socioeconomic (SES) status tend to have less empathy for people with mental illness and view them as being more blameworthy. This is partly because they see mental illness as caused by internal, controllable factors that a person has not taken responsibility for (Foster and O'Mealey 2022). This attitude might also result from people with higher SES status being less compassionate, less generous, and less other-oriented in general. Consequently, people with higher SES status display greater mental illness stigma, increased endorsement of punishment for people with mental illness when they behave badly, and increased desire for social distancing from them. In contrast, people with lower SES status are sometimes more empathetic toward people with mental illness. This is theorized to be the case because they rely more on other people's emotions to gain knowledge about a situation, are able to assess others' emotional cues more accurately, and are more other-oriented in general, for example giving to charity at higher rates than people of higher socioeconomic class. This emotional attunement and other-orientation leads to greater empathy toward and less stigma against people with mental illness (Foster, Elischberger, and Hill 2018, 139–140).

People of higher socioeconomic class can still develop understanding and empathy toward people with mental illness. Knowledge about mental illness increases empathy and decreases stigma, so that people who have greater knowledge of mental illness, regardless of socioeconomic status, display greater empathy and less stigma (Foster, Elischberger, and Hill 2018, 139–140).

Education levels also affect stigma, especially anticipated stigma and self-stigma. The more educated someone with mental illness is, the less hope they have regarding their illness and their future, and the more they anticipate stigma from others. This is especially true for people of color but also applies to White people (Rodwin et al. 2025). It is not evident why higher education levels are associated with greater anticipated stigma, but perhaps more educated people are more aware of their social environment and more self-conscious about how they are perceived by others, leading to greater fear of judgment.

How Gender Contributes to Stigma

Gender also affects stigma. Women tend to have more positive attitudes toward and less stigma against people with mental illness compared to men, in part because they have greater internal motivation to respond without prejudice (Hampton and Sharp 2014). Crosby Hipes proposes that women are less stigmatizing of others in general than men are because they have less power in society and are more empathetic toward those who have less power and are more vulnerable. She finds that, in the context of gender, "the more power one has, the more negative are one's evaluations of others" (Hipes 2016, 12–13). In addition, people who have mental illnesses associated with the gender that is opposite to their own are often judged more harshly than people who have mental illnesses associated with their own gender, and they are viewed especially poorly when they work in jobs or otherwise occupy social roles that violate gender norms (Koenig and Eagly 2014).

Compounding Stigmas

Many people who experience mental illness stigma are subject to other types of stigma as well, making their stigma compounded. Experiencing double or triple stigma from multiple sources can worsen people's experience, subjecting them to greater prejudice and discrimination, both worse in severity and worse in scope, covering a larger portion of their daily lives. Other types of stigma that many people with mental illness face include stigma related to race, gender, and homelessness.

People of color who have mental illness experience compounding stigma due to their different identities, experiencing both mental illness stigma and racial stigma (Misra et al. 2021). In one study, 93 percent of people of color who have mental illness said they have experienced mental health discrimination, and 73 percent said they have experienced racial discrimination. About 49 percent of these surveyed individuals have experienced discrimination from mental health professionals in both racial and mental illness contexts (O'Hara 2014, 40). Latinx people and Black Americans are especially wary of seeking mental health treatment because of anticipated stigma from mental health professionals in relation to both their race and their mental illness (Fripp and Carlson 2017; Shannon et al. 2022). In some racial and ethnic groups, people with mental illness prefer to seek help from informal sources like religious leaders, friends, and family, rather than formal providers like mental health professionals because they fear being stigmatized (Deng and Rose 2025).

In addition to experiencing more stigma, including mental illness stigma within their racial group, and both racial stigma and mental illness stigma within the general public, people of color also experience more barriers to accessing mental healthcare services. These barriers include practical barriers of being unable to afford services, to transport oneself to services, or to follow through on commitments to using services, as well as racial (and sometimes mental illness) stigma from mental healthcare professionals and stigma from one's social group about using mental healthcare services (Misra et al. 2021). Stigma related to marginalized race causes both mental and physical health problems, including increased stress, anxiety, and depression, as well as avoidance of services to help the homeless where people might fear being stigmatized for their race by service providers (Weisz and Quinn 2018).

Perceived dangerousness is tied to stigma and desire for social distancing. One social group that is particularly linked to perceived dangerousness is Black homeless men with mental illness. Black homeless individuals are perceived as more dangerous than White homeless individuals, and men who are homeless are perceived as more dangerous than women who are homeless (Markowitz and Syverson 2021). Black men with mental illness are also viewed as more dangerous than White men with mental illness or women with mental illness (Balogun-Mwangi, DeTore, and Russinova 2023). Black men who are homeless are perceived as particularly dangerous (Markowitz and Syverson 2021), and this is more so when they are believed to also have mental illness. Latino men are also subject to compounding stigmas of being perceived as dangerous in relation to both their Latino identity and their mental illness identity, especially when they have certain mental illnesses such as PTSD (Mann and Antunez 2024). The more dangerous a person is perceived to be, the more that people desire social distance from them (Markowitz and Syverson 2021).

In the workplace, Black people with mental illness feel patronized and rejected by others and are both held to lower expectations than others and assigned less demanding tasks. Their style of dress can contribute to the prejudice and discrimination aimed at them as it may be associated with criminality, incompetence, and lack of reliability. Black women who are known to have mental illness experience more sexual harassment in the workplace, and Black men who are known to have mental illness are perceived as more dangerous, threatening, and incompetent than others (Balogun-Mwangi, DeTore, and Russinova 2023, 9–10).

Men in general are perceived as more dangerous, and women are typically seen as more vulnerable, passive, and easy to take advantage of. Women with mental illness are more likely to be sexually exploited or harassed, or victims of violence. Both men and women experience gender stigma when they violate the gender norms of their sex, and their stigma is compounded when they have other stigmatized identities such as homelessness or substance use.

Women who are homeless and have mental or physical illness and use substances experience multiple intersecting and compounding stigmas based on these different identity statuses. Some of the ways they negotiate stigma include avoiding substance use themselves, avoiding other people who use substances, participating in social networks and social activities, and forming more positive identities for themselves based on positive values such as religion or motherhood (Thoams and Menih 2022). Women experiencing homelessness have high rates of trauma, violence, and victimization (Benbow et al. 2019, 203; Schreiter et al. 2021, 401; Thoams and Menih 2022, 2975). They experience a lack of safety in almost all places, including in their family situations, in their housing and neighborhoods when they have housing, in shelters when they are unhoused, and in hospitals and the healthcare system more generally (Benbow et al. 2019, 206–208). Women who have children and who are homeless also experience stigma by service providers, which often comes in the form of greater surveillance of their actions and more judgment about their parenting styles and behavior (Benbow et al. 2019).

People who are members of the LGBTQ+ community commonly experience double stigma from both their sexual orientation/gender expression status and their mental illness status. Older adults are subject to both ageism and mental illness stigma. Young people experience more perceived stigma in education and workplace settings (Torales et al. 2023). This perception may come in part from a heightened fear of being judged that young people have as a result of growing up with social media, where participating in social media makes them more self-conscious about how they come across to others.

Low SES status and unemployment are other conditions that are highly stigmatized. People commonly blame those in poverty for their condition, attributing their poverty to bad choices they made and lack of volition and responsibility in getting themselves out of their state (Okamoto and Peterson 2022, 390). This judgment rests on an assumption of neoliberal values which regard individuals as responsible for their socioeconomic condition, both in a backward, causal sense that they caused their condition through their choices and actions, and in a forward, taking-responsibility sense that they are responsible for taking the action required for changing their circumstances. In the dominant narrative, people who are unemployed are seen as being personally responsible for their lack of employment rather than viewing unemployment as a structural problem with the way the economy is organized (Holmes and Burgess 2021).

Mental health problems and financial difficulties compound people's social exclusion and resulting social isolation (Haugland, Topor, and Friesinger 2024). When people are subject to both mental illness and economic hardships, they experience deeper and more persistent social exclusion, partly as a result of the intersecting stigma that accompanies both statuses. Financial distress and mental illness are both conditions that make a person feel like they do not belong to the community. Moreover, both are conditions that others feel discomfort with and even fear (especially if they worry that they themselves could be subject to these conditions as well). As a result, others often shun and avoid people with these conditions, possibly in part so they do not develop these conditions themselves through perceived contagion.

The status of homelessness is highly stigmatizing as well. There is just as much stigma for homelessness status as there is for mental illness status (Phelan et al. 1997), and when people have both identities, their stigma is compounded. Some studies have found that people blame homeless individuals for being in their condition (Phelan et al. 1997), while other studies have found that people have a more nuanced understanding of the causes of homelessness and are less likely to blame homeless individuals for their condition (cited in Markowitz and Syverson 2021, 920). However, many people who are homeless have other stigmatized identities such as being male, being of a marginalized race, being unemployed, or having a substance abuse problem that intersects with, and compound stigmas of homelessness and mental illness. Service providers who aid the homeless can be stigmatizing in how they deal with the people they serve, especially when they have multiple stigmatized identities. Moreover, the general public often subjects homeless individuals to social distancing, social rejection and avoidance, prejudice, and discrimination.

Homelessness stigma, coupled with mental illness stigma, creates many harms. The stigma of homelessness leads to increased psychological distress, impairments in physical health, and avoidance of services to help the homeless in an effort to avoid the stigma associated with being homeless and receiving aid (Weisz and Quinn 2018). People who are subject to the interlocking stigmas of mental illness and homelessness often reach a stage of resignation, where they experience ongoing feelings of depression, anxiety, stress, a sense of unworthiness and lack of intrinsic value, alienation from both self and others, hopelessness, and helplessness (Marinucci et al. 2023). This typically worsens their mental illness symptoms and hinders both mental health recovery and the ability to exit homelessness.

Conclusion

Many people with mental illness face stigma from multiple directions based on several aspects of their identity. Mental illness stigma is thus often compounded by other stigmas as well. People with mental illness have to navigate many intersecting social groups that they are part of to try to avoid or address existing stigmas. At the same time, these social groups can be important sources of belonging and support that help protect against stigma. They can also help restore the humanity of a person with mental illness who has been devalued by broader society. When people with mental illness are accepted as equal members of these social groups, they are able to participate in the moral, epistemic, and social activities of the group, allowing them to exercise their agency and autonomy as full human beings.

Whether stigma is projected onto others or internalized within the self, it begins with the awareness of negative stereotypes that can be endorsed and/or perpetuated, sometimes knowingly and sometimes unconsciously. Mental illness stigma is based on various negative stereotypes related to what it is to have poor mental health and to receive mental health treatment. The following chapter examines what stereotypes are, how negative stereotypes about mental illness are perpetuated, and how these harm people who have mental illness.

CHAPTER 2

The Wrongfulness of Stereotypes

Introduction

As explained in the Introduction to this book, stigma starts with labeling people as being a member of a social group that shares a culturally undesirable trait and stereotyping people based on that trait. When the trait is devalued and discredited, negative stereotypes are applied to people who have that trait (in contrast to when the trait is valued, and positive stereotypes are applied). Whether stereotypes are positive or negative, they consist of generalizations that assume that any person with the stereotyped trait has the characteristic, positive or negative, associated with the trait. Stereotypes create expectations of what the person with the trait is like, and the person is judged according to how they seem to carry out those expectations.

Stereotypes are wrongful and harmful in many ways. In this chapter, I argue that what makes stereotypes wrongful is that they overgeneralize about a person's experience in a way that prevents others from recognizing the nuance and complexity of the experience of the stereotyped person. They reduce a person's identity to the characteristic associated with the stereotyped trait. Negative stereotypes reduce a person's identity to a negative characteristic that marks them as being bad and inferior compared to others. Whether negative stereotypes are projected onto others or internalized within the self, they involve judging the person who is stereotyped as being bad and inferior, which invites and is seen to justify reactions like prejudice, social exclusion, and discrimination.

This chapter begins by explaining what stereotypes are and how they can be held consciously as endorsed beliefs or unconsciously as implicit bias. Stereotyping involves making several kinds of judgments, including judging the validity of the stereotype itself, assessing whether particular individuals have the stereotyped trait and the characteristics associated with the trait, and evaluating whether an individual is good or bad for having or not having that trait. Stereotypes can be wrongful for many reasons, including the fact that they are often fixed and unchangeable, that they are epistemically false, that they make the wrong normative judgment about a trait, and that they ignore the nuance and complexity of a person's experience. It is the latter reason, I argue, that makes stereotypes truly and consistently wrongful.

What Stereotypes Are

Stereotypes as Generalizations about Members of a Social Group

Stereotypes are generalizations that we make about a group of people based on some shared characteristic (such as race, gender, or ability status) that is thought to be associated with another shared characteristic (such as a personality trait, ability, or strength or weakness). Stereotypes are applied both to groups and to individuals (Beeghly 2021a, 548). Groups of people are stereotyped when the group is associated with a certain trait in a way that portrays the group as a collective unit. These associations can include, as Erin Beeghly says, "beliefs about typical group traits, mental imagery, or webs of associations with informational significance" (Beeghly 2021a, 548). Stereotypes often function as beliefs about groups of people, but they can also involve other types of associations that are not full-fledged beliefs. What matters to stereotyping in relation to groups of people is the generalization that is made that associates the group, and individual members of the group, with a certain characteristic.

Individuals are stereotyped when they are judged negatively based on their membership in social groups, which is determined by possessing certain characteristics that are associated with that group (Beeghly 2021a, 548). Social groups that are defined in part by the possession of certain traits have social salience in that membership in the group conditions many types of social interactions. Kasper Lippert-Rasmussen describes social salience as follows: "A group is socially salient if perceived membership of it is important to the structure of social interactions across a wide range of social contexts" (Lippert-Rasmussen 2014a, 30). Some groups that share a certain trait are not socially salient, for example people who have blue eyes. Having blue eyes does not help structure the ways a person can interact with others. Members of groups that are socially salient possess a shared trait that is seen as morally relevant to how the person acts and how they are treated by others, and thus seen as relevant in structuring social interaction. In relation to the social group *people with mental illness*, for example, mental illness is usually (in the Western world) seen as a trait that is morally relevant to how a person who possesses that trait acts and how they are treated by others, making the social group *people with mental illness* a socially salient group. Groups are thus socially salient when they are defined at least in part by traits that are deemed morally relevant to behavior and social interaction.

Social salience is determined in part by cultural context. What counts as morally relevant to how a person acts and is treated by others can vary across different cultures. For example, Ethan Watters points out that in Zanzibar, schizophrenia is seen as an impermanent experience caused by temporary spirit possession and is therefore not viewed as affecting a person's identity or agency in the strong and decisive way that it is in the United States (Watters 2010, 127–185). While in the United States schizophrenia is seen as a life-changing condition that impacts a person's experience and agency globally and potentially permanently, in Zanzibar schizophrenia is viewed as a temporary condition that can make a person unwell for a short duration of time while they are occupied by the spirit, but that a person can recover from it when the spirit leaves them. Schizophrenia thus has a different social salience in Zanzibar than it does in the United States, and, as a result, it is not as stigmatizing.

Individuals can be stereotyped based either on their actual membership in a certain social group or on *appearing* to be members of the social group. Even when people may not be members of a social group, if they resemble other members of the social group in certain ways and/or exhibit certain characteristics associated with the social group, they can be presumed to be members of it. This presumption is sufficient for a person to be stereotyped. Simply appearing to be a member of a relevant social group can inspire others to apply a generalization to them that is associated with the social group and to assess whether they fit that generalization and thus whether the stereotype applies to them.

It is important to distinguish here between two types of traits that are relevant to stereotyping. There is the trait that is shared, or presumed to be shared, by all members of the group and thus partly constitutes group membership, and there is the characteristic associated with that trait that forms a generalization about people who have that trait. I shall call the trait that is shared by all members of a social group and thus partly constitutes the group a "group trait," "shared trait," or "social trait," while I shall call the characteristic that is generalized about in relation to that group trait a "generalization," "stereotyped trait," or "stereotype."

For example, people with mental illness share the group trait of having mental illness, or at least as seeming either to oneself or to others as having mental illness. The trait of having or seeming to have mental illness is the trait that all people in the social group designated as *people with mental illness* share, and that partly constitutes what it is to be a member of the group *people with mental illness*. When people who share the trait of having mental illness are stereotyped, their trait of having mental illness is associated with another characteristic, such as incompetence, neediness, defectiveness, dangerousness, or incurability, and generalizations are made that link the group trait with the stereotyped trait. For example, when people with mental illness are stereotyped as being incompetent, needy, or dangerous, generalizations are made that all people who are members of the group *people with mental illness* are incompetent, needy, or dangerous. In other words, stereotyping makes a generalization that presumes that all people who share the group membership trait also share an associated characteristic that is the basis of the stereotype.

A stereotyped trait is not always a *trait* in the sense that it is a way of describing a person, such as incompetence, neediness, or dangerousness. Sometimes a stereotyped trait is a state that the person who is a member of the relevant social group is in, such as being homeless or in jail. Sometimes it is a behavior that is associated with a certain social group, such as taking medication or receiving mental healthcare. For the sake of convenience, I will refer to descriptions, states, and behaviors all as "traits" or "characteristics," but traits can be any of these kinds of associations that can be made in relation to being a member of a certain social group.

Stereotypes as Beliefs or Associations

Stereotypes often function as beliefs, though they do not always or necessarily comprise beliefs. When they function as beliefs, a person *believes* the generalization about the social group that is stereotyped and endorses this belief as true. In the act of stereotyping, they then *apply* the stereotype to a person who is a member of the social group that is stereotyped, judging whether that person fits the stereotype by sharing the trait that is associated with their group. Endorsing a stereotype as true leads to prejudice, or a negative bias against a person or group of people that typically consists of an affective reaction such as fear, disgust, pity, or blame. Stereotypes are thus inextricably linked to prejudice; when they are endorsed, they consist of a cognitive state (an agreed-upon belief) that creates the negative affective response of prejudice.

Stereotypes are often (though not always or necessarily) essentialist. Ulrike Rangel and Johannes Keller identify five aspects of essentialism in this context: the belief that members of a social group have a shared and consistent propensity to possess and express certain traits; the belief that social groups are uniform, where all members share the same relevant traits; the belief that ways of characterizing social groups are distinct and non-overlapping; the belief that group membership is firm and unalterable; and the use of group membership to make what one believes are reliable inferences about members of groups based on their group membership (Rangel and Keller 2011, 1059–1060; see also Beeghly 2021b, 44). When a person stereotypes someone based on their membership in a social group, they often activate these beliefs, viewing members of the social group in these essentialized ways.

These aspects of essentialism sometimes show up in how people think about individuals who have mental illness. When people with mental illness are stereotyped, they are regarded as having a shared and consistent propensity to possess and express traits such as incompetence, neediness, defectiveness, dangerousness, and/or incurability. All people with mental illness are often assumed to share these traits, except, of course, for exceptions. People with mental illness are seen as fundamentally different from people without mental illness, so while people with mental illness are regarded as incompetent, needy, defective, dangerous, or incurable, people without mental illness are usually regarded as not sharing the propensity for these traits. Sometimes people with mental illness are viewed as being unable to change their mental ability status and their group membership and so the application of the generalizations made about them appears to be fixed.

A person can be essentialist about a group in either a biological or a social sense, whether attributing essentialist characteristics to a group based on biological factors such as genetic makeup or based on social factors such as upbringing, education, and socialization (Beeghly 2021b, 44). People with mental illness are often essentialized in both ways. Either they are seen as having something inherently defective with them due to biological factors such as faulty brain chemistry or flawed genetics, or they are seen as having inadequate coping skills due to social factors such as experienced trauma or poor upbringing. In either case, they are often seen as having something fundamentally wrong with them in an essentialist way by virtue of having mental illness.

Stereotypes do not always function as beliefs. Lawrence Blum argues that they are a mental construct that associates a group of people who share the same trait with a generalizable characteristic to which the trait is seen to be connected (Blum 2004). A person can make the association between the group of people and the generalization based on the stereotyped trait without endorsing the stereotype as a belief. In other words, stereotypes do not have to be endorsed in order to function in the cultural milieu. The mere association between a group and the generalization made about the shared trait is enough for a person to employ this association through their actions. A person can even actively repudiate a stereotype and nonetheless act on it because they retain the association between the group and the generalization. In fact, the association is often held unconsciously through implicit bias.

Implicit bias is a negative view that one has toward people who belong to a certain social group where members share a trait that helps constitute that group membership (Holroyd 2012; Holroyd, Scaife, and Stafford 2017; Levy 2017). This negative view is often held unconsciously at a level below one's awareness and self-knowledge, though sometimes people can be conscious of their biases (Nanay 2021, 330). Actions that stem from implicit bias tend to be performed without a person having any intention of directing negative energy and attention toward someone based on their group membership and often without any awareness that the action involves having a pejorative attitude or having negative effects on a person (Holroyd, Scaife, and Stafford 2017, 4). Implicit bias arises from internalizing culturally dominant negative stereotypes about certain groups of people based on a stereotyped trait they seem to share as a result of their social grouping.

Even though implicit bias occurs under the radar of one's conscious awareness, it can have devastating effects on people by treating some people differently from others, and it can shape interactions with others in profound ways that a person may be completely unaware of. For instance, implicit bias affects one's demeanor and bearing with people and influences what language a person uses, their body posture and positioning, and their interpretations of emotions and emotional reactions to others. It shapes how other people's speech and action are interpreted and responded to.[1] Implicit bias also conditions interactions, behaviors, and decisions such as how one treats and talks to certain groups of people, how one makes candidate selections for various positions, how one makes hiring decisions,

[1] Implicit bias can even affect how we hear the sounds of utterances. Because of stereotypes we hold of African-American men being dangerous and violent, we might interpret a Black man's utterance as "I have a gun" when in fact he said "I want some gum" (Peet 2017).

how one chooses tenants, how one interacts with clients and patients, and how one conducts evaluations of others such as formal employee evaluations or more informal judgments. Often, implicit bias is responsible for microaggressions, which are speech and behavior that subtly reflect prejudice and damage the person targeted.

Implicit bias is implicit not only in the sense that it often occurs unconsciously, below one's awareness and knowledge of it, but also in the sense that it may be outside of a person's control (Holroyd, Scaife, and Stafford 2017, 5; Levy 2017). If such bias is something a person has little or no control over, then even if they recognize they have this bias, they may have no ability to change their mental states and behavior to correct the bias. Jules Holyrod describes implicit bias as "automatic" in at least two ways: it does not arise through conscious effort but rather through unconscious processes one has little or no awareness of, and it cannot be revised based on reasons and evidence; in other words, it cannot be changed through rational deliberation and conscious effort (Holroyd 2012).

Often, implicit bias consists of prejudice that is unendorsed and can even go against beliefs that a person consciously endorses (Holroyd, Scaife, and Stafford 2017, 5–6). When the attitudes and behavior that a person consciously endorses and adopts conflict with their implicitly biased behavior, this can cause cognitive dissonance. The discomfort of cognitive dissonance can motivate a person to try to address the dissonance by trying to change the biased behavior so it is better aligned with their consciously held views. When bias is automatic, however, it may not be easily correctable through rational deliberation and conscious effort. In such cases, a person may have to work to publicly disavow and even compensate for biases that they have little or no control over.

Implicit bias results from holding negative stereotypes toward a certain group of people based on generalizations made about a social trait they share. These stereotypes can consist of beliefs, associations, or other belief-like states. While conscious bias stems from negative stereotypes that a person endorses as beliefs they take to be true, unconscious bias may stem not from fully formed or endorsed beliefs but rather from other belief-like states such as associations, "in-between beliefs," "patchy endorsements," or "aliefs." Eric Schwitzgebel suggests that unconscious bias arises from "in-between beliefs" that are not full-fledged beliefs in that they do not track one's attitudes and behaviors consistently (Schwitzgebel 2010), while Neil Levy argues that implicit bias comes from "patchy endorsements" that have propositional structure, similar to beliefs, but are unlike beliefs in that they are not responsive to reasons the way full-fledged beliefs are. As a result, they cannot be integrated into a person's other mental states well, and so arguably it may not be appropriate to attribute them to the person the same way beliefs can be attributed to the person who holds them (Levy 2017, 13–14; Levy 2015).

An alternative way to understand the belief-like states that give rise to implicit bias is what Tamar Szabó Gendler calls "aliefs" (Gendler 2008). Aliefs are states that have three kinds of content which are co-activated together by what happens in one's environment and what one is called to respond to; these consist of representational content,[2] affective states, and behavior (which can be understood in this context as readying of motor responses) (Gendler 2008; see also Albahari 2014). Aliefs are associative, automatic, a-rational, and activated by a person's environment. When Blum argues that stereotypes are often better understood as associations rather than beliefs, he probably means they are something like aliefs in that they associate two constructs together, they occur automatically within a person's mind, they occur outside of a person's rational consciousness or deliberation, and they are activated by what happens to a person in their environment. These associations help shape a person's behavior and interactions with others in the ways described here.

When people act based on stereotypes held through implicit bias, they perpetuate these stereotypes, often unwittingly. Blum says, "Someone 'expresses' a stereotype if she engages in any public behavior, or has a conscious mental response, that draws on a stereotype (and the individual does not repudiate the stereotype)" (Blum 2004, 269). When the bias is held unconsciously and unknowingly, the perpetuation of the stereotype occurs without the person's conscious awareness that they are engaging in this perpetuation; they might in fact be quite surprised that they are doing so. But, as Blum points out, actions do not have to be consciously based on stereotypes to perpetuate them; action that unconsciously sustains and supports stereotypes perpetuates them as well. "Expressing" a stereotype is sufficient to perpetuate the stereotype in society; endorsement of it is not necessary (Blum 2004, 270).

When people express stereotypes, they consciously or unconsciously draw upon and refer to generalizations that are part of the cultural and historical milieu. Stereotypes are *cultural* constructs that circulate throughout the culture by many members of the society. They do this through the perpetuation of the associations that exist between the generalized trait and the social trait. This perpetuation of

[2] Bence Nanay argues that the representational content of implicit biases should be understood as mental imagery or perceptual processing (Nanay 2021).

stereotypes occurs in how people act and respond to people in the stereotyped group as well as how they act and respond to others in relation to people in the stereotyped group.

The Judgments Involved in Stereotyping

While a stereotype can be seen as a belief that members of a certain social group are a certain way, or as an association between a certain social group and a particular generalization that is linked to that social group, the act of stereotyping involves making a number of judgments about what to believe or associate and how beliefs and associations apply to particular social groups and individuals. Judgments thus lead to (though they do not themselves consist of) beliefs and associations. Stereotyping involves making generalizations about people who appear to share a morally relevant trait, but it also involves making judgments about whether particular individuals fit these generalizations and how these individuals should be regarded as a result.

Judgments are made at both the individual level and the group level. When stereotypes are under consideration, social groups are judged in relation to the stereotyped trait that is associated with their group. The group is viewed as good or bad depending on what value is assigned to the stereotyped trait associated with the group. When individuals are stereotyped, there is always an evaluative component to the stereotyping, where individuals are assessed in relation to the generalization made of their social group. Individuals are viewed as good or bad depending on how they conform to or violate stereotypes associated with their social group.

Stereotyping involves making at least three kinds of judgments. One kind is the judgments made regarding the social construct that consists of the stereotype under consideration. Here, social groups are judged as being appropriate objects of certain generalizations made of them, and stereotypes are evaluated as being more or less true and endorsed when they appear to be accurate. The other two kinds of judgments regard the individuals who are being stereotyped. Individuals are judged as to whether they are members of the relevant social group and whether the generalization associated with the social group applies to them, and they are evaluated as being good or bad depending on how the generalization is viewed and whether the stereotype does or does not apply.

The Judgment of Whether a Stereotype Is True or Not

Stereotypes do not always consist of beliefs; sometimes they consist of associations made unconsciously through implicit bias. Nevertheless, stereotypes can usually function as beliefs in that they can be endorsed as true or not. One kind of judgment that a person makes in the process of stereotyping is a judgment involving endorsing a stereotype or not: believing whether it is true or not and thus putting their stamp of epistemic approval or disapproval on it. An example involves deciding whether to endorse the association between mental illness and other characteristics such as incompetence, neediness, defectiveness, dangerousness, and incurability.

This judgment is supposed to be made descriptively, based on facts, though it is not always; often it smuggles in values and biases. Ideally, a person examines the facts around whether people with a certain group trait are likely to have a certain associated characteristic that can be generalized about them *because* they have that group trait. In other words, the association between the group trait and the stereotyped trait must be a causal one: A person must have the stereotyped trait on account of having the group trait. If this is seen to be the case, then the generalization that people who have the relevant social trait tend to have the associated stereotyped trait is a belief that can be endorsed.

Sometimes stereotypes are endorsed as true, and sometimes they are repudiated at a conscious level, though they may be so insidious that the implicit bias involved with the associated generalization is resistant to evidence and reasons. In making a judgment about whether a stereotype is true, cognitive biases often play a role in skewing how a person perceives and interprets evidence, falsely coloring the person's judgment. Shared reality bias, illusory correlation bias, and in-group favoritism bias can all shape whether a person accepts a stereotype as true, consciously or unconsciously.

In shared reality bias, people conform their own beliefs and views to culturally dominant beliefs so that their own way of understanding the world fits the culturally accepted framework (Anderson 2012, 170; Anderson 210, 46). In the process, they ignore and fail to seek out other possible interpretations. When a culture has negative stereotypes about people with mental illness, such as that they are inherently defective, incompetent, needy, dangerous, or incurable, people adopt these views as their own and close their minds to alternative understandings of mental illness that challenge such stereotypes.

The illusory correlation bias invites people to see correlations between different phenomena even when none actually exist (Anderson 2010, 46; Balzan et al. 2013, 320). In the context of stereotypes, this allows people to adopt stereotypes that certain phenomena are common, regular, and normal when in fact they are unusual and abnormal. Such phenomena seem common, regular, and normal because they are vivid images that circulate in the culture through media portrayals and may be shared repeatedly so they appear to be more common than they are. Even though such images can become popular ways of seeing a phenomenon, this popularity may belie the fact that the phenomenon is actually rare and abnormal. Images of people with mental illness perpetuating violence are prime examples of images that are vivid and commonly shared, and so presumed to be common and normalized, when in fact they are rare and abnormal. People adopt the stereotype that people with mental illness are violent based on the popularity and vividness of such images, but this stereotype does not reflect the truth, that very few people with mental illness actually perpetuate violence. Only about 12 percent of people who have certain mental disorders have violent behaviors, and this accounts for only 4 percent of all violent acts (Appelbaum 2020, 345; Corrigan and Cooper 2005, 171). The stereotype does not reflect reality.

In-group favoritism bias makes people have a positive bias for members of their own social group and a negative bias against members of other social groups (Anderson 2010, 46). This bias encourages people to identify with and feel a sense of belonging with people like themselves and to regard people who are dissimilar to them as profoundly different and "other" and to see them as not belonging to one's community. A person may regard people who have mental illness as being fundamentally different from themselves as a way of reifying their own identity as *not* mentally ill and affirming their own mental health status as healthy, rational, and sane in opposition to the "othered" group. In this way, people define themselves in relation to what they are not, thus affirming their identity through separation of those who are healthy and normal from those who are not (Gosselin 2018, 82; Ottati, Bodenhausen, and Newman 2005, 107–108). This reifies their sense of themselves as "normal" and their sense that people unlike themselves, such as those with mental illness, are abnormal, which in turn encourages them to see people with mental illness as having something "wrong" with them and thus as being defective.

All of these biases – and others – contribute to a person endorsing stereotypes about groups of people, such as people with mental illness, that do not accurately depict the reality and lived experience of the people being stereotyped. Instead, people maintain stereotypes based on biased perceptions and understandings of the evidence of what the people who are being stereotyped are like. The judgment of whether a stereotype is true or not often rests on biased beliefs about what people are like that do not correspond with reality. Such a judgment can result in a false and unjustified belief or an unsupported association, but one that is nonetheless taken as knowledge by the person who holds it and that is thus treated as knowledge in their assessments and resulting behavior.

The Judgment of Whether a Person Has a Morally Relevant Social Trait and Whether They Conform to a Stereotype or Not

A second kind of judgment made in the process of stereotyping is an ostensibly descriptive judgment that assesses whether certain facts apply to a person. In making this judgment, one examines whether the person being judged has the traits that are being attributed to them – the social trait that places them in a certain social group and the stereotyped trait that is associated with that social group – and makes a judgment about whether they do or do not possess those traits. The judgments about whether a person is a member of a certain social group and whether they possess the associated stereotyped characteristic are ostensibly descriptive because they seem to be judgments of fact. A person either is or is not a member of a given social group, and they either do or do not conform to the associated stereotype. However, such judgments actually smuggle in normative assessments in terms of how a person's appearance and behavior are interpreted.

In the context of mental illness, the judgment of whether a person has mental illness frequently involves assessing a person's appearance and behavior to see if these seem to indicate mental illness. People with mental illness transmit many different signals that come from their mental illness and get generalized as being attributes of people with mental illness. Because these signals or signs are associated with mental illness, their presence in someone can be interpreted as signifying mental illness in that person (Corrigan and Kleinlein 2005, 13–14). Some of the signs that suggest mental illness include behavior that could qualify as psychiatric symptoms, inappropriate social interactions, disheveled appearance, odd demeanor, and labels. Behavior that could qualify as psychiatric symptoms includes inappropriate emotions, emotion-reading, and emotional response; strange behavior that does not fit the situation; peculiar uses of language or speech that do not fit the situation; and talking to oneself.

Inappropriate social interactions can include behavior that is not deemed socially appropriate, such as poor eye contact; overly close or far bodily positioning; threatening, off-putting, or isolating body language; and speech that does not appropriately respond to what others are saying or doing. Unkempt appearance, poor hygiene, and poor self-care often result from mental illness symptoms and can be further signs that suggest mental illness. Being labeled mentally ill or having a known diagnosis of a diagnosable mental disorder also signifies mental illness. Any of these signs can be interpreted by someone who knows to associate these attributes with mental illness as indicating mental illness and thus be used to make the judgment that someone has mental illness.

However, there are many factors that can color how someone assesses a person's appearance and behavior, and different interpretations can be made depending on what background assumptions, biases, and attitudes an evaluator has. For example, a person who appears reluctant to engage with others could be seen as being shifty and untrustworthy, and consequently viewed as immoral, seeming to conform to the stereotype of people with mental illness as immoral, unreliable, and untrustworthy. Or the person who seems reluctant to engage with others could be viewed as distracted, perhaps seen as evidence of hearing voices or otherwise being preoccupied with one's internal states, and consequently viewed as incompetent, another stereotype of mental illness. Or the same behavior can be seen as evidence of shyness, which may not be regarded as indicating mental illness at all. Which interpretation a person has of such behavior depends on factors such as background beliefs about mental illness; whether the evaluator views the person as being part of their community or as an "other"; and whether the evaluator holds more benevolent, suspicious, or judgmental attitudes toward the person. These can all affect whether the evaluator interprets evidence (the person's appearance and behavior) as supporting or undermining the judgment that they have mental illness.

Stereotypes about other social groups can also interact with mental illness stereotypes to affect how a person is perceived and whether they are thought to have mental illness or not. Stereotypes of Black men as hostile, dangerous, and violent overlap with stereotypes about people with mental illness as dangerous and violent, so Black men are more likely to be seen as having mental illness and their behavior is more likely to be interpreted as hostile and threatening. For example, Jonathan Metzl gives a case study of how residents of a psychiatric hospital in the 1970s largely consisted of Black men, who were regarded as having symptoms related to dangerousness and hostility, in comparison with White men in the hospital, whose symptoms were seen as more benign, such as social withdrawal and apathy (Metzl 2009). Black men's behavior was more likely to be seen as symptomatic of mental illness, and in particular of dangerous mental illness.

Moreover, whether a person is viewed as conforming to a stereotype associated with the social group they are judged to be members of, or not, depends on whether the stereotype is endorsed or not. When an evaluator endorses a stereotype as true, they are more likely to interpret a person with the relevant social trait as fitting the stereotype because they expect to see the stereotype instantiated. This constitutes confirmation bias. A person's behavior is more likely to be seen as conforming to the stereotype when the stereotype is believed to be true – unless the person is seen as an exception, in which case their exceptionality is viewed as confirming that the generalization is largely true. For example, when a person with mental illness seeks help, they may be viewed as dependent and needy when the evaluator assumes that the stereotype of people with mental illness as dependent and needy is true.

When an evaluator refutes a stereotype as false, on the other hand, they are more likely to interpret a person – who may or may not have the relevant social trait – as not fitting the stereotype, because they do not expect to see manifestations of the stereotype. This is because confirmation bias is not invoked. The person's behavior is more likely to be interpreted in alternative ways that do not support the stereotype. For example, when an evaluator rejects the stereotype of people with mental illness as dependent and needy, help-seeking behavior by people with mental illness could be seen positively as self-care or as an instantiation of the interdependence and interconnectedness of the relational self.

Furthermore, the attitudes held toward the person being judged affect how their behavior is interpreted. When the evaluator views the person negatively, such as with suspicion, fear, hostility, or blame, their behavior is more likely to be seen as fitting a negative stereotype. When the evaluator views the person positively, such as with benevolence, compassion, or empathy, their behavior is more likely to be seen as not conforming to negative stereotypes. Whether the person being evaluated is seen as a member of the community also affects interpretation of their behavior, where people who are seen as members of the shared community are viewed in a more positive light while people who are seen as outsiders or outcasts are viewed more negatively. How behavior is interpreted, and whether it is seen as indicating membership in a certain social group (such as *people with mental illness*) or as fitting a stereotype or not, depends on what background beliefs an evaluator holds, what attitudes they have toward the person they are evaluating, and whether they endorse a given stereotype or not.

The Judgment of Whether the Stereotyped Characteristic Is Good or Bad to Have, and Whether a Person Who Is Seen to Have the Stereotyped Characteristic (or Not) Is Viewed as Good or Bad

A third kind of judgment that a person makes in the process of stereotyping is a normative judgment about whether the stereotyped characteristic – the generalization that is being made about people who share a certain group trait – is good or bad to possess, and, relatedly, whether a person who appears to have that stereotyped characteristic (or not) is good or bad for having it (or not). Whether a stereotyped characteristic is seen as good or bad, and whether people are judged as good or bad for possessing it, depends on cultural and historical factors that shape how the characteristic is perceived and valued.

Cultural factors such as ideals and values of the culture frame the way a characteristic or behavior is seen and what value is placed upon it. For example, consider the stereotype that associates having mental illness with being weak for receiving mental healthcare treatment. In a culture that values independence, self-reliance, and solving one's problems oneself, seeking help for one's difficulties can be looked down upon (Courtney, Dial, and Yanos 2024; Lui, Sagar-Ouriaghli, and Brown 2024). In such a culture, receiving mental healthcare to help deal with mental health problems can be viewed as a negative thing because it is seen as a way of outsourcing responsibility to deal with one's problems onto someone else rather than taking responsibility to do so oneself. In this context, receiving mental healthcare can be understood as a violation of cultural values that a person internalizes as a member of that culture.

Such a cultural context also invites people to demean individuals who have other traits often associated with mental illness, for example being incompetent or needy. In a culture that values independence and solving one's problems oneself, being needy or incompetent is seen as a gross violation of such values and can prevent people from valuing help-seeking (Lui, Sagar-Ouriaghli, and Brown 2024). When people who have mental illness are assumed to be needy or incompetent, they are seen as violating important cultural values, making neediness and incompetence highly negative stereotypes that one would not want to embody.

Receiving mental healthcare can be viewed negatively in other cultural contexts as well. In a culture that distrusts industries that make money off of helping people deal with problems, perhaps because they are seen as vultures preying on the vulnerable, receiving mental healthcare can be seen as negative because it involves paying people to do something that one should be able to do oneself (or at least find a cheaper alternative such as asking friends for help). In this cultural context, distrust of mental healthcare services is similar to distrust of addiction rehabilitative services and weight loss services which spring up as industries designed to help people achieve wellness for a profit (Courtwright 2002, 108–111).

In a culture that values interdependence and compassion and normalizes seeking and attaining help to deal with difficulties, on the other hand, receiving mental healthcare can be viewed as an asset rather than a weakness. It can indicate strength in knowing when to ask for help from others and having faith and trust in others to be able to help one successfully. Receiving mental healthcare can be viewed as supportive of and supported by values such as strength in asking for help, having trust and faith in others' capabilities, and being interdependent. In a culture that values these ideals, other stereotyped traits such as incompetence and neediness may also be viewed differently. Incompetence and neediness can be seen as vulnerabilities that must be protected and as occasions to receive compassion, help, and goodwill of others. Cultural values thus can promote stereotypes that are viewed as either good or bad depending on what the culture values and how the stereotyped characteristic fits within that value context.

Historical factors also shape how a characteristic or behavior is perceived and valued, and how a characteristic is viewed can change over time in different historical contexts. In other words, the normative content of generalizations can change over time and in different cultural circumstances. For example, in the past, in the United States, receiving mental health treatment was seen as bad because treatment was seen as a bad thing to participate in, and engaging in treatment was seen as indicating an inherent weakness. When people internalize the value of self-reliance and the belief that one should have control over one's mental states and behavior, as were common values and beliefs in the past, they tend to avoid seeking professional help (Lui, Sagar-Ouriaghli, and Brown 2024).

Today, however, most people in the United States do not believe that receiving mental health treatment is inherently bad or signals an inherent weakness. In fact, many people who receive mental health treatment do not have severe mental illness, or even any mental illness at all, as they engage in mental health treatment to *improve* their mental health without necessarily having something defective or faulty with their mental health. As a result, receiving mental health treatment is no longer associated with the inherent weakness and inferiority that having mental illness was often associated with. The characteristic of receiving mental health treatment is often associated with the social trait of having mental illness, but it is not *always* associated with the social group; it can reflect a desire to improve one's mental health status rather than fixing a mental illness flaw.

Moreover, even for people who do have a mental illness, attitudes toward seeking treatment have shifted in recent years to be more positive toward treatment (Abdelmonaem, Osman, and Karim 2024). People's views toward mental health professionals are more positive, and the value of mental health treatment is more recognized, increasing the desire to seek professional help (Angermeyer et al. 2017; Onyeukwu and Donahue 2024, 432). People also have better mental health literacy due to the pervasiveness of mental health education campaigns, which also improves help-seeking (Donovan et al. 2024; Lui, Sagar-Ouriaghli, and Brown 2024). Improving one's mental health status, whatever that status is, can be seen as a positive thing insofar as fostering one's own well-being is good for a person. Rather than connoting that a person has an inherent defect, as it did in the past, therefore, receiving mental healthcare now sometimes has the connotation of improving mental health and general well-being. In this way, receiving mental health treatment can be viewed positively as an effort to enhance well-being, functioning, and fulfillment and can be valued whether a person receives treatment because of mental illness or not.

The judgment of whether a person is good or bad for fitting a stereotype or not gives rise to certain kinds of actions and reactive attitudes. Someone is more likely to engage in positive actions such as helping when they perceive the person being judged as good. They are more likely to engage in negative actions such as being dismissive or mean to the person when they perceive the person as being bad. People apply negative emotions and reactive attitudes such as blame, shame, anger, disgust, fear, and resistance to people whom they judge as bad, while they apply positive emotions and reactive attitudes such as approval, pride, satisfaction, happiness, liking, and welcoming to people whom they judge as good. What kinds of action a person engages in and what reactive attitudes they hold depend on the normative judgment they make about a person based on whether a stereotype seems to apply or not.

How a person is judged according to stereotypes also affects how others interact with them. Whether the person is viewed as like oneself or as radically different, and whether they are welcomed as a member of the community that belongs or whether they are outcasted or shunned are all shaped by the judgment of whether a person is good or bad. Such a judgment affects whether the person is taken seriously and granted credibility or whether they are discredited, ignored, dismissed, or discounted, and whether they are seen as trustworthy or untrustworthy. This judgment also determines whether the person is ascribed positive or negative motivations, and whether their beliefs, desires, and goals are viewed sympathetically or suspiciously. These in turn shape how one responds to and interacts with the person being judged.

Judgments about whether a person fits a stereotype and is consequently good or bad are made in the moment of encounter, but judgments about whether a stereotype is true or not are typically made in the past, based on cultural and historical biases that shape understanding. Thus, some of these judgments are retrieved from one's bank of what they consider to be knowledge, while others are formulated in real time. People do not usually reflect on when a stereotype can be appropriately used and what makes stereotypes typically wrongful, but if they are going to use stereotypes in developing an understanding of others, they should.

The Wrongfulness of Stereotypes

Stereotypes are typically wrongful, for many reasons. They are epistemically wrongful in that they resist counterevidence and counter-reasons and are often maintained with rigidity. While some philosophers regard stereotypes as epistemically wrongful by being false or misleading generalizations that do not match reality, I contend that stereotypes do sometimes contain a grain of truth. Nevertheless, even when they are based on some truth, they overgeneralize in a way that ignores the complexity and nuance of people's lived experience and thus are both epistemically problematic and morally disrespectful and unjust. While some stereotypes have a descriptive component that can be evaluated epistemically as more or less accurate, many stereotypes also have a normative component that unjustly and problematically assumes the badness of the characteristic that is stereotyped. Moreover, some stereotypes are wholly normative in that the stereotyped characteristic is regarded as inherently bad-making or inferior-making to those who are seen to possess it, making these stereotypes especially morally problematic and unjust.

The Rigidity of Stereotypes

One of the properties of stereotypes that make them bad, according to Blum, is their rigidity or fixedness (Blum 2004, 256). Blum argues that stereotypes are rigid because they are resistant to evidence: When a person endorses a stereotype, they affirm the belief underlying the stereotype despite whatever evidence there is that would disprove it. They seek out evidence that seems to support the stereotype and either ignore or explain away evidence that undermines it (Blum 2004, 261). Rather than revising their belief or changing the association they make between a social trait and a stereotyped trait in light of evidence that undermines that belief or association, a person upholds their belief and revises their interpretation of the evidence to fit their belief. In this way, stereotypes are rigid and fixed: static characterizations of groups of people that do not match reality and are not revisable in the face of evidence. Being resistant to evidence is epistemically wrong because it leads to unjustified belief and unsupported associations, where a person is unable to justify their belief or association because they are unable to take into account properly the evidence that is before them.

The rigidity of stereotypes applies both to consciously held beliefs that are maintained even in the face of conflicting evidence as well as to unconsciously held associations that constitute belief-like states such as aliefs and other types of associations. The implicit bias that people internalize is often as firmly and stably held as fixed conscious beliefs; even when a person becomes aware of their unconsciously formed association and is motivated to try to change it so it better fits their consciously held values of wanting to treat everyone equally, they are often not able to effectively change their bias. Such bias is often resistant to rational deliberation, counter-reasons, and counterevidence. The stereotypes that underlie both conscious beliefs and unconscious bias are frequently fixed and not revisable based on evidence.

Seeing someone who shares the stereotyped trait but does not fit the stereotype does not change people's practice of stereotyping. If someone seems not to fit the stereotype, they are viewed as an *exception* that proves the rule, rather than as evidence that the rule is in error. People hold onto their stereotyped views regardless of the evidence before them and explain away evidence of disconfirmation as exceptional. People who seem to be exceptions to negative stereotypes are viewed positively in contrast with those who are seen to embody the rule; people who are viewed as exceptions to a negative stereotype are often lauded as exemplary, because they are seen as having overcome the negative aspect that is the object of the stereotype.

This ability to overcome the negative aspect that is stereotyped is viewed as a quality that improves a person and can lead to the person being viewed as *better than* other people in their group because of their ability to overcome the negative aspect. This can make a person seem more similar to "normal" people who are not in the stereotyped social group than to other members of that group, leading them to be more welcomed and have a greater sense of belonging in the larger community. This exceptionality only reaffirms what appears to be the truth of the stereotype, however, suggesting that those who are unexceptional – who are most people in the stereotyped social group – fit the negative stereotype that is applied to them and thus seem to confirm the truthfulness of that stereotype.

The Epistemic Accuracy of Stereotypes

Another property of stereotypes that makes them bad, according to Blum, is that – he claims – they are always false (Blum 2004, 256–257). He examines and dismisses the view of stereotypes that they contain a "kernel of truth" to them. The fact that *some* people in a group share the trait that is the basis for the stereotype does not make it true of the group; within *any* group of people, we are liable to find that *some* members of the group share any number of traits, but this does not indicate anything about the group itself. Blum acknowledges that a group could have a greater statistical prevalence of having some experience or trait that is the basis of the stereotype, but he says that this is irrelevant for two reasons. First, people do not generally know the statistical prevalence of an experience or trait for any given group and certainly do not base their beliefs on this, so it really does not matter what the statistical prevalence of a trait is in relation to how we form beliefs about it. Second, stereotypes are generalizations that are felt to apply to most or all members of a group. Having a greater statistical prevalence of an experience or trait does not make it generalizable in the same way; even if more members of the group share that trait or experience than people in the general population, or in other social groups, the trait or experience is not shared by most or all members of the group. In these ways, Blum contends strongly that stereotypes are always false.

Erin Beeghly disagrees with Blum, arguing that stereotypes are generalizations that can be more or less true (Beeghly 2021b). Drawing on Ariel Cohen's work, she contends that generalizations can be accurate to different degrees. A generalization is what Cohen calls an "absolute generic" when it is true that a simple majority (more than 50 percent) of a social group has the trait in question and is a "relative generic" when it is true that members of a social group possess the trait at a higher rate than people who are not members of the social group (Beeghly 2021b, 41; Cohen 1999). I do not know what percentage of people who have severe mental illness are dependent on others and on government and nonprofit systems to get their basic needs met or to aid in the ability to participate in daily activities, so I do not know whether the generalization that people with mental illness are dependent is an absolute generic. However, it most certainly is at least a relative generic, as more people with mental illness are dependent in these ways to a greater degree, statistically speaking, than the general population. Thus, there is a sense in which this stereotype is epistemically accurate, at least for many people with mental illness.

Moreover, the stereotype that people with mental illness are incompetent is based on a similar fact that people with mental illness, to a greater degree than the general population, struggle with difficulties functioning in various parts of their lives. Thus, this stereotype, too, is at least a relative generic. Because mental illness is in part *defined* by difficulties with functioning, it is probably an absolute generic as well. The most recent edition of the *Diagnostic and Statistical Manual of Mental Disorders* (DSM-V) defines a mental disorder as "a syndrome characterized by clinically significant disturbance in an individual's cognition, emotion regulation, or behavior that reflects a dysfunction in the psychological, biological, or developmental process underlying mental functioning. Mental disorders are usually associated with significant distress or disability in social, occupational, or other important activities" (American Psychiatric Association 2013, 20). The stereotype that people with mental illness are incompetent is rooted in part in the fact that people with mental illness typically have difficulty functioning in various life domains, a fact that partly constitutes what it is to have mental illness. Thus, this stereotype is, to some extent and in some ways, epistemically true.

Other stereotypes of people with mental illness may not be either absolute generics or relative generics and can thus be seen as epistemically false. The stereotype that people with mental illness have a permanent and incurable condition is probably not an absolute or even a relative generic. I do not know what percentage of people this includes, but many people with mental illness do recover from their illness, either episodically, as their illness goes into remission for periods of time, or fully; for many people, their illness is not permanent. The stereotype that people with mental illness are dangerous and violent is neither an absolute nor a relative generic. Studies indicate that only about 12 percent of people with mental illness engage in violent behaviors, and only 7 percent when people who have comorbid substance abuse are not counted. While this is higher than the 2 percent of the general population that engages in violent behaviors, it accounts for only about 4 percent of all violent acts (Appelbaum 2020, 345; Corrigan and Cooper 2005, 171). The belief that people with mental illness are dangerous overgeneralizes by a lot, making a small subset of the population the image that is seen to represent all in the population.

Thus, while some stereotypes may be false in the ways that Blum describes – for example, the stereotype that people with mental illness are largely violent is certainly false – some other stereotypes may be, to some extent, true. As I just noted, the stereotype that people with mental illness are incompetent is based on the fact that many people with mental illness do struggle with competence in at least some areas of their lives, and the stereotype that people with mental illness are dependent and needy is based on the fact that many people with mental illness do depend on other people, on the mental healthcare system, and on other systems and government agencies such as disability, housing, and employment services to help them get their basic needs met and to help them with daily functioning. What is problematic about these stereotypes is not that they are false – there is indeed a "kernel of truth" in them – but rather that they generalize too much without acknowledging the nuance and complexity of experience and identity.

The stereotype that most people with mental illness are incompetent generalizes too much in the sense that most people who struggle only struggle in some domains but are quite competent in other domains. Very few are globally incompetent – incompetent in all areas of their life – though global incompetence is a possibility for a very few. The stereotype that people with mental illness are incompetent is based on the fact that people with mental illness struggle with some aspect of mental functioning in at least some areas, but it generalizes too much to suggest that most or all people with mental illness are globally incompetent. The global incompetence is false, but it is indeed based on a "kernel of truth." What makes the stereotype that people with mental illness are incompetent problematic is not that it is wholly false, but rather that it generalizes too much.

Similarly, the stereotype that people with mental illness are necessarily dependent on others and on various government and nonprofit systems also has a grain of truth in it yet also generalizes too much. Many people with mental illness are dependent on others in a variety of ways, including emotionally, financially, and in terms of being able to get their basic needs met and being able to participate in daily activities on their own. Some people are not more dependent in these ways than the general population, but many are. All people with mental illness who receive treatment for their illness are dependent on the mental health system for ongoing treatment, and many people with mental illness are also dependent on hospitals, disability services, and housing services. This is especially true for people with severe mental illness such as schizophrenia, bipolar disorder, and major depression. The problem with the stereotype that people with mental illness are dependent is not that it is false – in many cases it is at least to some extent, in some ways, true – but that it generalizes too much and misses the nuance and complexity of a person's experience and identity. It is true that many people are dependent in many ways, but they are also capable of being independent and interdependent in other ways, and this generalization elides this fact.

Assessing the Normative Evaluation Inherent to Many Stereotypes

Moreover, stereotypes that people with mental illness are incompetent and dependent assume that being incompetent and being dependent are bad things. Many stereotypes are not merely descriptive generalizations that can be more or less true; they also include a normative component in which the stereotyped trait has normative content and is viewed as positive or negative. When the stereotyped trait is viewed as positive, having the trait appears good and makes the person possessing the trait praiseworthy. When the stereotyped trait is viewed as negative, having this trait is seen as bad, making the person a candidate for blaming, shaming, and shunning. For many stereotypes that have this normative component, when a person is stereotyped, they are judged positively or negatively according to whether they have the relevant trait. When people with mental illness are stereotyped as being incompetent or dependent, it is assumed that being incompetent or dependent is a bad trait to have, and people are judged negatively when they are seen as fitting the stereotype.

One of the aspects of stereotypes that is problematic, therefore, is that they contain a normative evaluation that may be unwarranted. Is being insufficiently competent or overly dependent really a bad thing? If it is not, then these stereotypes would not be seen as negative. If being incompetent or dependent is not bad, it should not matter whether a person is insufficiently competent or overly dependent. They may indeed fit the stereotype of being incompetent or dependent, but that shouldn't make a moral difference. The proper response to the assertion that many people with mental illness are necessarily incompetent or dependent should then be: So what?

Stereotypes of incompetence and dependence are negative because these are regarded as undesirable traits to have. If these traits were viewed more positively, or at least in a morally neutral way, the generalizations that many people with mental illness struggle with competence and functioning and that they are dependent on others and on institutions and systems for caretaking, to get their basic needs met, and to participate in daily activities would not be seen as negative or problematic. It would be a descriptive generalization rather than a normative one. However, we live in a world (at least in the United States in the twenty-first century) where being incompetent and dependent are seen as bad things, so being regarded as incompetent or dependent is seen as bad and as warranting blaming, shaming, and shunning. Many stereotypes are wrongful not only because they are epistemically problematic as fixed beliefs and biases that generalize too much but also because they are problematic in how they evaluate people normatively. They make assumptions about what characteristics are good or bad to have that may be unjustified.

Purely Normative Stereotypes and Prejudice

In addition, some stereotypes cannot be assessed as to whether they are absolute or relative generics or not because they are purely normative and do not reflect any description that can be epistemically evaluated. Stereotypes that individuals who are members of certain social groups are lesser – for example, abnormal, defective, having lower moral status, or not being fully human – are purely normative and have no value-neutral description that can be assessed as to whether they are true or false. Such stereotypes do not have epistemic content that can be evaluated; they are moral statements that can only be evaluated through a moral lens, based on moral criteria. For example, the stereotype that people with mental illness are defective is a value statement that can only be evaluated from a moral standpoint. Whether a person's illness should be considered a "defect" is a value question about how to regard people with mental illness in relation to other people. The stereotype that people with mental illness are fundamentally different from other people and thus abnormal is a value judgment that views illness as an inferior-making trait that makes people who have this trait inferior to people who do not have it. This inferiority is often used to justify thinking of people with mental illness as less than human and thus dehumanizing them.

People who are members of multiple minority groups are often subject to intersecting stereotypes that dehumanize them in a plurality of ways. For example, Black people may be seen as inherently inferior because of fully normative stereotypes that denigrate Blackness, so that Black people with mental illness are seen as inferior both because of their Blackness and because of their mental illness. In fact, the two stereotypes could be conjoined so that Black people are viewed as more likely to have mental illness *because of* their inherent inferiority.

Other people of color and people with physical disabilities can be multiply dehumanized in similar ways when their other social groups are also seen as inferior-making. The stereotype of indigenous people as lazy alcoholics denigrates indigenous individuals as necessarily inferior, while the stereotype of Latinx people as menial workers supports viewing Latinx people in necessarily pejorative ways; these stereotypes can intersect with and reinforce purely normative negative stereotypes about mental illness that mark people as inherently inferior both because of their illness and because of their race. Similarly, when physical disabilities are viewed as necessarily inferior-making, they also reinforce purely normative stereotypes about mental illness that dehumanize people who have both mental and physical disabilities in multiple ways. In addition, inferior-making social identities can be seen as causing other inferior-making social identities. For example, indigenous people, Latinx people, and people with physical disabilities might all be assumed to have mental illness *because* they are already presumed to be inferior due to their race or ability status. Purely normative negative stereotypes that represent people as necessarily inferior reinforce each other to dehumanize people in a plurality of ways, especially when people are members of multiple social groups that are subject to such stereotypes.

Stereotypes that are purely normative most easily support prejudice. Prejudice is a negative affective response to people who have a social trait that is seen as inherently negative. Corrigan and Kleinlein argue that prejudice stems from stereotypes, where people agree with and thus endorse the negative beliefs involved with negative stereotypes and have a negative emotional reaction as a result, such as anger, fear, disgust, or a desire to avoid people who have the relevant social trait (Corrigan and Kleinlein 2005, 16–17). I would argue that often prejudice is not based on negative *beliefs* that have semantic content but rather based on negative associations people make regarding a social trait. The negative association may be something general, like associating a social trait with abnormality, defectiveness, subhumanness, inferiority, and lower moral status, but the negative associations do not necessarily have content that goes beyond these vague characteristics.

Nonetheless, these negative associations can be understood as stereotypes that are purely normative because they are inherently bad-making and inferior-making. In other words, the characteristic associated with the social trait is necessarily and always negative, and having the social trait necessarily and always marks a person as inherently bad or inferior. When a person holds purely normative stereotypes that associate a social trait with abnormality, defectiveness, subhumanness, inferiority, and lower moral status, they almost always develop an affective reaction such as anger, fear, disgust, or wanting to avoid people who have that social trait. Thus, negative stereotypes with purely normative content almost always produce prejudice in people, whether the person is aware of it (where the stereotype is a conscious belief one endorses) or not (where the stereotype is more of an association than a belief, and where it can be held at an unconscious level as implicit bias).

Assessing the Descriptive and Normative Components of a Stereotype

Some stereotypes can be separated into descriptive and normative components, while others are always normative. The generalization that people with mental illness are incompetent can be viewed descriptively, in a value-neutral way, as an epistemic description of people's problems with functioning, although it often also involves a normative component that judges incompetence as bad. The generalization that people with mental illness are dependent can be seen descriptively, in a value-neutral way, as an epistemic description of people's dependence on others and on systems, but again it often also includes a normative aspect that judges dependence as a bad trait to have. If we remove the normative component of these generalizations, so that we do not assume that possessing these traits is bad, we are left with a mere (value-neutral) description of some people's situation. In such a case, people who possess these traits would not be judged poorly for possessing them. Having these traits would be a simple fact about people's experience, neither good nor bad, just is.

Some stereotypes are more value-neutral than others, at least in some cultural contexts. For example, the generalization that people with mental illness need mental healthcare treatment is largely true for many people with mental illness and is neither a good nor a bad trait to have; it just is.

The generalization that people with mental illness are suffering is also largely true for many with mental illness. The fact that they are suffering is negative in the sense that it is unfortunate for them to experience, but it does not make the person a bad person for having this trait. In other words, the fact that they are suffering is not a bad-making characteristic; it is value-neutral in terms of how individuals who have this trait are viewed. Value-neutral stereotypes can be seen as descriptions of people's experience and identity that can be epistemically evaluated as to how true they are, and they do not affect how individuals who are subject to these stereotypes are viewed (as good or bad).

The generalizations that people with mental illness are permanently defective, and that they are fundamentally different from other people, on the other hand, are intrinsically value-laden. We cannot separate out a descriptive component that does not include a normative component, because the normative evaluation is built into the generalization. Having a defect is, by definition, bad. Being fundamentally different from other people or abnormal is usually considered bad in contexts where a person is being compared to others, because usually there is an assumption that normalcy is what is desirable. If we changed the way we thought about mental illness so that we no longer saw it as a defect, or so that we no longer saw it as making the person who possesses it fundamentally different from other people, abnormal, or inferior to other people, then these generalizations would cease to exist. If mental illness is not a defect, then generalizing people with mental illness as being defective no longer makes sense. If mental illness does not make someone fundamentally different from others, then there would be no generalization that people with mental illness are fundamentally different from others.

For stereotypes that are necessarily and intrinsically normative, removing the normative component destroys the stereotype. For these stereotypes, there is no descriptive component that can be separated out from the normative component and preserved as an issue of fact. What is needed to change these stereotypes, then, is to remove the normative evaluation they involve, to change the way we think about what it means to have a social trait such as having mental illness.

Beeghly proposes that we understand stereotypes as fundamentally descriptive, rather than normative, as a way of forming and using concepts to generalize an understanding of people (or things) (Beeghly 2021b). In this context, assessing a stereotype involves an epistemic assessment of how accurate the generalization is. However, this approach only works for certain stereotypes, the ones that have a clear descriptive component separable from any normative component. For stereotypes that are inherently normative, on the other hand, a different approach is necessary. Then assessment is not a matter of epistemic determination of accuracy of the generalization, but rather a moral determination of the ethics and justness involved with how people are regarded in the generalization.

Conclusion

Stereotypes are generalizations made about people who are members of social groups that share a socially salient trait, where that trait is associated with another characteristic that forms the generalization. Stereotypes can be consciously believed and endorsed, or they can be held unknowingly and unconsciously as implicit bias. Either way, they are hard to challenge and resist, because they are often automatic, a-rational, and activated by one's social environment. Stereotyping someone involves making judgments about the veracity of the stereotype, evaluating whether a person has the relevant social trait and the generalized characteristic that is associated with that trait, and assessing how good or bad a person is based on having or not having the stereotyped trait. Stereotypes are wrongful for many reasons, including their rigidity, their frequent falsity, and the normative values they smuggle in, but most of all because they ignore the nuance and complexity of a person's experience.

In the next four chapters, I examine some of the harms that stereotypes, and the prejudice and discrimination they support, create for people who are stereotyped. Stereotypes are not merely wrongful in that they are held tenaciously despite counterevidence and counter-reasons, in that they overgeneralize in ways that ignore the complexity and nuance of lived experience, and in that those with normative components unjustly attribute bad-making and inferior-making characteristics to people. They are also extremely harmful. In the rest of this book, I detail ways that people with mental illness in particular are harmed through stereotyping. The next chapter examines harms to the self, identity, and agency that result from stereotyping.

CHAPTER 3

How Stereotypes Shape Self and Identity

Introduction

Stereotypes shape self and identity in several ways. One way is through creating what Elizabeth Anderson calls "expressive harm," which is harm that occurs through the expression and perpetuation of stereotypes and stigma independently of whether there is any intention behind them (Anderson 2010, 44–66). Stereotypes circulate within a culture regardless of whether people endorse or apply them, and they have the power to control the thoughts, feelings, and behavior of both people who can be stereotyped and people who can subject others to stereotypes. In always hovering in the social atmosphere, they color social interactions in unavoidable ways, influencing the agency and identity of those who can be stereotyped.

Stereotypes also cause stereotype threat, in which individuals who are aware that they can be judged and stereotyped by others work hard to resist conforming to a stereotype, but in the process impair their performance in ways that end up making it seem like they fit the stereotype (Steele 1997; Steele and Aronson 1995). In trying to be better than what a stereotype would indicate of them, an individual experiences stress and anxiety that leads to inadequate performance that makes them look insufficiently competent, also causing exacerbation of mental illness symptoms and emotional reactions like hostility and aggression. All of these can reinforce negative stereotypes about what it is like to have a mental illness. This affects a person's self-concept and sense of their personal identity, especially when they internalize the stereotypes they are trying to resist.

In addition, stereotypes affect self and identity through the internalization of what Natalie Stoljar calls "oppressive social scripts" (Stoljar 2015). Social scripts attribute mental states and behaviors to a person based on dominant, and often oppressive, conceptions of what people with a stereotyped trait are like. In this way, they determine how a person's past behavior is interpreted and create expectations for how a person will behave in the future. Through this process, an individual is judged according to how they conform to the oppressive social script, and others' interactions with them are based on this. Oppressive social scripts shape how a person perceives their own agency and identity and influence how they interact with others.

Stereotypes thus impact a person's sense of self and personal identity in several ways. They have the power to control the mental states, behavior, and social interactions both of the people who are in a position to stereotype others and of the people who can be stereotyped. This chapter shows some of the ways that stereotypes shape self and identity through expressive harm, stereotype threat, and the internalization of oppressive social scripts.

Expressive Harm

Stigma invokes negative stereotypes about people who share a social trait that circulate within a culture and have a historical and social context. When people stigmatize others, they draw upon these cultural stereotypes in forming their attitudes and behavior toward those whom they are stigmatizing. When people with a stigmatized trait internalize stigma, they incorporate into their self-understanding and self-concept cultural stereotypes that circulate in the culture regardless of particular individuals' endorsement of them. Cultural stereotypes are transmitted through social activity, often without intention, and persist regardless of whether particular individuals endorse them.

Drawing on Patricia Hill Collins's work, Erin Beeghly argues that many stereotypes are not just psychological constructs that help organize how one understands other people, but they are also *cultural* constructs that have the power to control people. Collins views stereotypes as controlling images or ideas that affect people's beliefs, values, and desires, as well as thoughts, emotions, and behavior (Collins 2000, 69–96). Even stereotypes about positive traits are controlling in that they aim, in Beeghly's words, "to direct how people are seen and treated in society, hence to control or regulate the nature of social relations" (Beeghly 2021b, 53). Regardless of the degree of accuracy a stereotype has, stereotypes maintain their power to control people both when they are consciously used and when they are unconsciously perpetuated or referred to. Even their mere existence is enough to control people.

Stereotypes have the power to control because they are expressed even when they are unendorsed. Stereotypes are expressed when a person thinks, feels, or behaves in a way that draws on stereotypes, whether or not the person endorses or even is aware of the stereotype (Blum 2004, 269–270). A person can be unaware of holding a stereotype through implicit bias, or they can even be unaware of the existence of the stereotype. Yet, when they think, feel, or act in a way that draws on and reflects the stereotype, they express and perpetuate it regardless of their intentions or consciousness around it. In other words, an act can express stigma even if that is not the intention, simply because it can be interpreted that way (Anderson 2010, 49). As Elizabeth Anderson says, "[i]deas can be practically engaged without being consciously endorsed" (Anderson 2010, 49).

There are multiple ways that stereotypes control mental states, behavior, and social interactions. Insofar as they are pervasive and persistent, they occupy a space in people's minds that they cannot avoid or ignore, manipulating people's thoughts and feelings. In doing so, they exert pressure on a person to make the person respond to them. Even as they force people to act in certain ways in relation to them, they also create constraints on people's actions, especially the actions of those who can be subject to stereotypes. They remove certain options from action in the sense that people are not free to act as if they are unaware of stereotypes, because the stereotypes are ever-present and make people constantly aware of them. They also channel people's actions in specific ways, forcing them into certain types of reactions. Through this, they exert the power to manipulate people's thoughts and feelings and coerce people to act in certain ways and not others, impacting their behavior and how they interact with others.

Stereotypes have the power to control the mental states, behavior, and social interactions of both others who do the stereotyping and the person who is stereotyped themselves. In other chapters, I examine the way stereotypes control others' thoughts, feelings, and behavior through social exclusion (Chapter 5) and discrimination (Chapter 6). In this chapter and in Chapter 4, I show some of the ways stereotypes control the mental states and behavior of people who are stereotyped. Chapter 4 shows how stigma and stereotypes shape a person's self-concept, self-worth, and self-regarding traits, including self-respect, self-trust, self-confidence, and self-efficacy. This chapter shows how stereotypes affect mental states and behavior by always being in the atmosphere and thus governing social interactions, by fostering internalized stigma and anticipated stigma, by causing stereotype threat, and by imposing the internalization of oppressive social scripts that shape identity and experience.

The Power of Stereotypes' Mere Presence: The Dance of Acknowledgment

Part of how stereotypes create expressive harm is through the way they maintain negative images of people that control how people think, feel, and act. Stereotypes serve as "default images" that help shape interactions between people, even when people do not endorse these stereotypes as true (Anderson 2010, 53). Usually individuals are aware of the existence of certain stereotypes that apply either to themselves or to the people they are interacting with, and they have to dance around the stereotype, either willfully trying to ignore it even as it influences what they do or publicly acknowledging it and either disavowing it or proclaiming that it does not apply to the current situation. Even when an individual is unaware of existing stereotypes, the person they are interacting with who could be subject to the stereotype tends to be fully aware of its existence and is obligated to respond to its presence, making the individual have to respond in turn even if they were previously unaware of the governing stereotype.

Anderson gives an example of the dance that an individual capable of stereotyping and a person who could be stereotyped have to do in acknowledging a relevant, existing stereotype, even when neither person endorses it (Anderson 2010, 53). She writes of a Black man who offers to help her with car trouble and reassures her that he is not going to rob her; in turn, she reassures him that she believes and trusts him. He has to publicly acknowledge the existence of the stereotype of Black people as criminals and show that it does not apply to the given situation; she has to acknowledge the stereotype as well and publicly disavow it, at least in this circumstance. Neither can ignore the stereotype without automatically invoking it.

In the context of mental illness, a person with mental illness might have to acknowledge that they can be perceived in a certain way – for example, as "crazy," as incompetent, or as dangerous – and then assure the person they are interacting with that they are not like that. The person interacting with them would have to acknowledge the existence of this stereotype and publicly disavow it or assure the person they believe it does not apply to them. If the person with mental illness does not acknowledge the stereotype, they are more likely to be seen as conforming to it. If the individual they are interacting with does not acknowledge the stereotype, their behavior will typically perpetuate it as they interact with the person with mental illness in a way that assumes that the stereotype fits the person.

In these scenarios, the stereotype helps govern how the individuals interact with and respond to each other, requiring them to take notice of the stereotype and requiring them to react to it in a publicly acceptable way. The individual who can be stereotyped is harmed through this process, even when the person they are interacting with disavows the stereotype, because their behavior is always colored by the possibility of that stereotype being applied to them. Anderson says of the Black man who offered to help her with her car troubles: "The harm consists in the fact that he walks under a cloud of suspicion in unstructured encounters with strangers. To gain access to cooperative interactions, he must assume the burdens of dispelling this cloud, or protecting and proving his innocence of imagined crimes" (Anderson 2010, 53).

As Anderson notes, stereotyping is not simply a process of thinking about certain groups of people in certain ways, making cognitive generalizations about their experience and identity. Rather, stereotypes have a "public standing" that makes them publicly known and available, something that must be referenced and acknowledged even when it is not taken to be true. They have a cultural existence and carry social meanings that pervade the way people think about and interact with others even when they are not activated by an individual. As a default image, they become part of the background conditions that shape behavior and social interaction, often in subtle ways that people are not even aware of. They shape the behavior and social interaction not only of the people who have the power to stereotype others but also of the individuals who could be stereotyped.

The power of stereotypes is thus far-reaching. Stereotypes can be perpetuated either through overt behaviors that are morally and behaviorally discriminating, where a person, institution, or practice knowingly treats some people differently based on a trait they share, or through more understated behaviors such as microaggressions that result from implicit bias. In other words, people can perpetuate stereotypes by consciously endorsing and holding them or by unconsciously acting in ways that treat certain people differently through the subtlty of microaggressions. The expressive harm of stereotypes goes much deeper than this, however. Even when people disavow stereotypes and eliminate their implicit bias, stereotypes still have the power to govern their social interactions because they persist in the atmosphere independently of particular individuals' endorsement of them. Even when stereotypes do not morally or behaviorally discriminate against people, they still have the power to control mental states, behavior, and social engagement.

Internalized Stigma and Anticipated Stigma

The expressive power of stereotypes typically causes individuals who can be stereotyped to internalize stereotypes that may apply to them simply through the power stereotypes have when they merely exist, even when they are not consciously perpetuated or activated. Through the cultural circulation of stereotypes, an individual sees how stereotypes can apply to people who have a certain social trait and recognizes that since they have that social trait, these stereotypes could be applied to them, too. Internalization of stereotypes does not require endorsing stereotypes as true or even believing that they do apply to oneself; it only requires that one sees how they *could* be applied to oneself depending on how they are judged by others. The mere existence of a stereotype serves as an open invitation for people who share the social trait that is stereotyped to be judged by others. As soon as a person recognizes the possibility of being judged by others in relation to the existing stereotype, they are subject to the power of the stereotype in shaping their mental states, behavior, social interaction, identity, agency, and experience. In this way, the person internalizes stigma through what I call "attendant stigma."

Seeing how a stereotype could apply to oneself entails that a person recognizes themselves as being able to be seen as having a social trait that can be stereotyped. This could involve self-labeling, when a person recognizes themselves as a member of the social group that is stereotyped, but this is not necessary. A person only has to be aware that *others* may label them as being a member of the stereotyped group. Even when people do not self-label as having a devalued social trait like mental illness or self-identify as a member of a socially salient social group such as *people with mental illness*, they are often aware of how they may be perceived by others and so often aware of the possibility that they can be judged in relation to the stereotype. (This requires that people have at least some self-awareness and insight that allow them to be self-aware of how they are perceived by others. Sometimes mental illness clouds self-awareness and makes it so that a person lacks insight, in which case a person might not be aware of how they may be perceived by others. However, many people with mental illness do have such awareness.)

The expressive harm of stereotypes frequently leads to anticipated stigma, where people expect to be stereotyped and stigmatized against, and they anticipate being subject to prejudice and discrimination. When a person realizes that a stereotype could be applied to them, they might assume that the stereotype *will be* applied to them, and they might prepare themselves for how they want to react to that stigma so that they are ready to respond when the moment calls for it. Anticipated stigma is another way that stereotypes control people, by making a person mentally engage in a certain way in preparation for how they will react to stigma.

Anticipated stigma can lead to different reactions. This includes self-stigma, or applying stereotypes and stigma to oneself, perhaps in order to exercise power by stigmatizing oneself before others do, or perhaps to align one's own views and reactions with those of others. Another reaction is outcasting oneself by withdrawing socially from interactions with others, again often as an exercise of power to outcast oneself before others can outcast one, where outcasting is a typical response to stigmatized individuals. Other reactions include getting defensive about stigma being applied to oneself. Sometimes people respond to anticipated stigma in more self-affirming ways by resisting it through self-disclosure of having the stigmatized trait and engaging in activism that aims at creating pride in one's identity rather than shame. Whether positively or negatively, anticipated stigma shapes the thoughts, feelings, behaviors, and actions of people who expect to be stigmatized by others.

Stereotype Threat

Another way that stereotypes control the mental states and behavior (and thus the sense of self and identity) of people who may be stereotyped is through stereotype threat. Stereotype threat is the potential to be seen as confirming a negative stereotype about a group of which one is a member based on the behavior a person engages in when they are aware that they could be judged by that stereotype (Steele 1997, 614; Steele and Aronson 1995, 797). A person does not themselves need to endorse the negative stereotype as being true or as applying to themselves in order to be subject to stereotype threat; it is sufficient that a person is aware that *other people* could endorse the negative stereotype as true and believe that it would apply to the person if they appeared or acted in ways that seem to correspond with the stereotype. What subjects a person to stereotype threat is an awareness of the potential for other people to judge oneself based on other people's belief in or at least perpetuation of the stereotype.

Stereotype threat makes stereotypes self-fulfilling in that awareness of a stereotype can cause a person to act or appear in ways that seem to affirm the stereotype. In this way, stereotype threat is another way that stereotypes control people's mental states and behaviors: the awareness of the potential for other people's judgment causes psychological effects in the person that affect the person's thoughts, feelings, and actions. Studies have shown that when Black children are primed to be aware of the negative stereotype that Black people are less intelligent than White people, they underperform on academic tasks; and when girls are primed to be aware of the negative stereotype that girls/women are less capable at math than boys/men, they underperform on math tests (Steele 1997; Steele and Aronson 1995). In some trials, children were not primed to be aware of the negative stereotype, but awareness of stereotype was nonetheless triggered when researchers did something to remind the children of their race (for Black children) or gender (for girls) (McKinnon 2014, 860; Steele and Aronson 1995, 808). Priming of the stereotype was thus not necessary; all that was necessary was a trigger that awoke children's awareness of a stereotype that they had already internalized. When Black children and girls were made aware of the relevant negative stereotypes, this awareness seemed to shape their cognitive activity and lead to worse performance.

Stereotypes can be self-confirming when stereotype awareness causes the underperformance that constitutes the stereotype. Being aware that other people may judge a person to be less intelligent based on their race can cause psychological responses, including anxiety, stress, and self-consciousness, which can lead to decreased confidence, cognitive mistakes, and disengagement from the task (Steele and Aronson 1995, 809). On the other hand, such awareness can lead to overworking cognitive capacities as an attempt to compensate for the potential of being seen to underperform, which can tire and stress a person out and lead to errors and fatigue, thus leading to the underperformance that the person is trying to avoid (Freeman 2017, 646; Steele and Aronson 1995, 809). A person who experiences these psychological and cognitive processes is no less intelligent than others, but they are subject to stresses that others who are not targets of stereotype threat are not subject to. Being aware that they may confirm a negative stereotype instigates psychological and cognitive processes that lead to the very thing – the confirmation of a stereotype – that the person is trying to evade.

What triggers the underperformance involved with stereotype threat is the self-consciousness people have of being judged. One study showed that stereotype threat had a worse effect on people's performance when it was triggered by a self-evaluation of appearing to confirm a negative stereotype compared to a public evaluation where others might believe the person fit the stereotype (Desombre et al. 2019). This was due to how self-conscious people are about how they appear to others. When people believe they can be judged negatively by others, this activates the response of stress and anxiety that can lead to underperformance.

People with mental illness may experience stereotype threat with respect to negative stereotypes such as the idea that people with mental illness are incompetent, dependent on others, inherently defective, fundamentally different than others, or lesser compared to others. When a person is aware that other people might view them as incompetent, dependent, defective, fundamentally different, or lesser, regardless of how they view themselves, this can make them self-conscious about how they act and how they appear to others. This self-consciousness can cause or exacerbate anxiety, creating a level of stress. Self-consciousness, anxiety, and stress can be crippling, causing problems in reasoning ability, emotional regulation, and volitional control. The attempt to compensate by trying to be more competent in various areas can lead to overwork of one's capabilities, which can lead to mistakes, underperformance, lack of confidence, and disengagement. These effects can make a person seem less competent or defective in cognitive and other kinds of activity so that they appear to confirm the stereotype of incompetence or defectiveness that they are self-conscious about wanting to avoid confirming.

Self-consciousness, anxiety, stress, and attempts to compensate can lead a person to seem to confirm other stereotypes about mental illness as well. In decreasing the performance of a person's abilities, it can make a person unable to take care of all of their own needs themselves so they need to rely on others for help, thus confirming the stereotype of dependence. Self-consciousness, anxiety, and stress can also exacerbate existing mental illness symptoms, making the person appear to be even more "abnormal" (as judged according to how mental illness symptoms manifest) and thus more different from others than they would appear without the awareness of the potential to be judged. This appearance of abnormality confirms the stereotype that people with mental illness are fundamentally different from and lesser than everyone else. Being aware that one can be judged as to whether one fits a stereotype thus can lead a person to behave or appear in ways that in fact seem to fit the stereotype.

Some studies have been done on the effects of stereotype threat on people with mental illness. In one study, when people who were diagnosed with schizophrenia believed that others knew of their diagnosis, the quality of their social interactions diminished, making them appear more socially inept. This in turn made them seem more like they fit negative stereotypes of people with schizophrenia as having low social functioning (Henry, Hippel, and Shapiro 2010). In another study, women and nonbinary people who were reminded of the gender disparity in diagnoses of depression, where women are twice as likely to get diagnosed with depression compared to men (Salk, Hyde, and Abramson 2017), had worse task performance compared to when they were not reminded of this (Tlachac et al. 2022). (The reminder of gender disparity in depression did not affect men's task performance.) Having received treatment for depression in the past also invoked stereotype threat for women, where women performed worse at tasks when they had reported this experience (Tlachac et al. 2022). Stereotype threat can easily worsen the performance of people with mental illness, which can make them seem to conform to stereotypes related to incompetence, neediness, dependence, defectiveness, inferiority, and even dangerousness.

Internalizing Oppressive Social Scripts

Stereotypes are often internalized in a deeper way than simply being conscious that one can be judged according to a stereotype and feeling forced to react to the stereotype and the potential judgment. The power stereotypes have over a person who can be stereotyped goes further than just affecting what they do, obligating them to act in certain ways, by also affecting who they are. Often a stereotype is internalized as part of a person's self-conception and identity in complicated ways that make simple rejection of the stereotype difficult. This integration into one's self-concept and identity can damage a person's autonomy in important ways.

Natalie Stoljar explains this form of internalization in terms not of stereotypes, which operate as beliefs or associations that generalize about a person based on a social trait they share, but rather in terms of social scripts (Stoljar 2015). Social scripts tell a story behind the stereotype about what the person has likely done, what their motivations probably are, how they tend to think and feel, and what they are likely to do in the future, all aspects of identity, agency, and experience that are presumed about a person based on their social trait. Stoljar shows how people who are subject to stereotypes are often forced to internalize oppressive social scripts in the sense that they have no choice but to accept that the script could be applied to themselves, where the script has power over how they see themselves and how they behave and interact with others (Stoljar 2015). They are forced to react to a script that is imposed on them not necessarily by other individuals (though it may be, either consciously or through implicit bias) but instead imposed on them through unavoidable cultural interpretations of who they are and what they feel, think, and do. Before we examine the implications of this internalization of social scripts, let us look more closely at what a social script is and how it differs from a stereotype.

Social Scripts

Stoljar does not define what a social script is, but I take it to be a story that describes who a person is based on some social characteristic the person has, where the description of the individual is framed by dominant narratives, including stereotypes, of what people who have that characteristic are like. An oppressive social script is not simply a stereotype, though stereotypes are embedded within scripts, as they are key concepts for the story to hinge on. A script goes beyond a stereotype to tell us a little bit more about the person, including providing a temporal context for who the person is. Part of the story embedded within a social script concerns what has led the person to be the way they are today, what motivates them, and what they are likely to do. A social script can be understood as a drawn-out stereotype, or as the story that goes along with the stereotype.

Scripts include attributions of mental states to the person whose identity is governed by the script, ascribing them with certain motivations, intentions, thoughts, feelings, and other mental and psychological characteristics. For example, a person with mental illness may be identified with characteristics such as having a chaotic, confused, or disorganized mind; being emotionally volatile; lacking self-control; or being dominated by obsessions or compulsions. A person with mental illness may be assumed to not understand the world around them, or other people's emotions and behavior, or even themselves, making them fundamentally different from other people. They may be assumed to have cognitive capacities that are slow and undeveloped, emotions that are unpredictable and untrustworthy, and motivations and behavior that are erratic and uncontrollable. They may be seen as irrational, disconnected from reality, excessively anxious, over- or under-emotional, or out of control. All of these ascriptions can become part of the oppressive social script involved with the stereotypes that people with mental illness are permanently defective, fundamentally different (abnormal) from, and lesser than other people.

Just as stereotypes are reductivist, a script often includes a reductivist ascription of causation of the person's behavior and mental states that is centered around the trait that is stereotyped within the script. It is not uncommon for other people to view the emotions, thoughts, responses, and behavior of a person with mental illness as essentially stemming from their illness, where their illness is the cause of their mental states and behavior. Whether a person feels anger, fear, sadness, or happiness, this is all attributed to their illness; whether they are disengaged or overly involved, whether they are inactive or hyperactive, whether they are inattentive or overly focused on something, this is all attributed to their illness as well. However they respond to others socially interacting with them, this is chalked up to their illness. We can call this attribution of mental states and behavior to the trait that the script centers around the *causal attribution of mental states and behavior* aspect of social scripts.

A script may also include a reductivist account of causation of the trait that the script is based on. For example, the idea that mental illness is essentially a brain disease – that it is caused by biological factors internal to the person – is part of the script that tells a story about people with mental illness that suggests they are permanently defective and fundamentally different from other people. In this script, the mental illness tends to be viewed reductively as a biological problem with a biological cause. Everything that a person says, does, thinks, feels, or responds with is presumed to be ultimately caused by this brain disease, since the person's mental states and behaviors are frequently seen to be caused by the illness, which in turn is seen to be caused by the brain disease. We can call this reductivist identification of the cause of the trait that is stereotyped the *causal account of the stereotyped trait*, and it often underpins the causal attribution of mental states and behavior as the ultimate underlying cause.

The causal account of the stereotyped trait has effects on how the person to which the script applies is viewed and helps support the stereotype(s) embedded within the social script. When a person with mental illness is presumed to have a brain disease, other people may interpret this as indicating that there is something fundamentally wrong with the person that cannot be changed and at best just be managed. Much or all of the person's behavior and responses become interpreted through the lens of having a brain disease, so that when the person struggles in any context, this tends to be seen as a result of the disease. As the problems the person faces become interpreted solely through this biological lens, so do the solutions. When a person is presumed to have a brain disease, the solution to their problems seems to be that they need to take medication to treat the disease. Their reasons why they should or shouldn't take medication become moot and are dismissed by others. They may become judged on whether or not they take medication and thus whether or not they are willing participants in the treatment of their disease, the solving of their problem. People who do not take medication, regardless of what their reasons for this are, may be judged negatively as being uncooperative, uninvested in their own welfare, and essentially immoral.

There are other reductivist accounts of the cause of mental illness. Sometimes mental illness is seen as the product of trauma, or of experiencing adverse events that one has insufficient resources to cope with. With these accounts of the cause of mental illness, other people tend to attribute all the mental states and behavior of the person with mental illness to problems coping with trauma or adversity. Most or all the struggles that a person experiences, even when these struggles do not seem to have anything to do with mental illness, may be nonetheless interpreted as a result of illness, as a result of inadequate coping skills. The idea that mental illness stems from an experience of trauma is embedded within a story about people with mental illness having insufficient coping skills compared to other people, who seem to be unable to learn and grow like other people, and who are thus made permanently defective and weak by their experience and condition. This may lead to a judgment of people with mental illness as being weak and weak-willed, having a character flaw, being unable or unwilling to "pull themselves up by their bootstraps," and not taking responsibility for themselves.

A social script also includes an account of what motivates the person to whom the script applies. In oppressive social scripts, these motivations are reductive. In the case of mental illness, these motivations are largely psychological and having to do with the person's pathology. Some of the motivations that are attributed to people with mental illness include delusions that don't make sense to other people, obsessions and compulsions that are irrational, distrust or excessive trust of other people, gullibility or suspiciousness and paranoia, ruminations and worries, and out of control emotions that are either excessive or inappropriate to the context. All of these can be seen to motivate a person's behavior, regardless of what reasons a person might give for their behavior.

Their own reasons for acting are typically ignored or dismissed, because their behavior is chalked up to mental illness symptoms. In this way, they frequently experience testimonial injustice with respect to their ability to make their reasons for acting known to others (Fricker 2007, 17–29).[1] As

[1] Havi Carel and Ian James Kidd analyze the way patients in the medical healthcare setting, including mental health patients, often experience testimonial injustice (Carel and Kidd 2014; Carel and Kidd 2017; Kidd and Carel 2017). The testimonial injustice experienced specifically by mental health

discussed in Chapter 1, people with mental illness are frequently subject to testimonial injustice where they are viewed as unreliable reporters due to their mental illness, and their testimony is often discounted or dismissed and not seen as worth taking seriously. The ability to make one's reasons for acting known to others is one such context where people with mental illness experience this testimonial injustice. They may try to give reasons for their behavior, trying to enact their agency as reason-givers, but be unable to make themselves heard because other people refuse to listen to them and take their reasons seriously. When other people presume that all of the person's behavior is motivated by mental illness symptoms, their reason-giving ability is thwarted, truncating both their agency and their autonomy.

Depending on which specific stereotypes are applied to the person with mental illness, they may be attributed with different specific motivations. For example, when a person with mental illness is viewed stereotypically as necessarily dependent on others, and on government or nonprofit systems such as the mental health system, the disability system, or the housing system, their behavior may be seen as them trying to take advantage of the system. This interpretation stems from a belief that people who are dependent on others, or on government or nonprofit systems, don't really *need* the assistance they are given and could do things on their own if they only tried hard enough. Underlying all of these attributions of motivation of a person's behavior is an element of judgment, where the person who is ascribed these qualities is judged poorly, and these qualities are seen as something one shouldn't have, or if they do have it, they should work to overcome it.

Another aspect of social scripts includes predictions about a person's behavior. These are usually tied up with specific stereotypes embedded within the script. For example, the mentally ill person who is seen as incompetent will seem like they will not initiate or work on projects, or if they do start a project they will not finish it. The mentally ill person who is seen as socially awkward will seem to try to engage in social interactions that are awkward and uncomfortable and ultimately fail. The mentally ill person who is seen as having a character flaw will seem to succumb to lack of willpower and motivation, making them lazy, apathetic, and lethargic. The mentally ill person who is seen as dangerous and violent will seem to engage in violent behaviors threatening the safety of those around them.

patients in a mental healthcare setting is explored in Rena Kurs and Alexander Grinshpoon (2018) and Abdi Sanati and Michalis Kyratsous (2015).

The mentally ill person who is seen as irrational and unpredictable will seem to engage in random and absurd behaviors that don't make sense, which may possibly be seen also as threatening and dangerous. Embedded within such scripts are judgments that these predicted behaviors are bad and undesirable and should be overcome. The fact that a person has not overcome them is seen as evidence of the person's weakness, defectiveness, incompetence, and abnormality, and consequently their badness as a person.

Social scripts thus often include accounts of causation of the person's mental states and behavior, accounts of the cause of the stereotyped trait, assumptions about what motivates a person to act as they do, and predictions about behavior. They are stories about who a person is, what has led them to be where they are today, why they act the way they do, and what they are likely to do in the future. Social scripts fill out details of stereotypes by providing a narrative about a person that follows from the stereotype. However, they are just as reductivist as stereotypes are, reducing a person's experience and identity to the social script and overlooking nuance and complexity about why people act the way they do, what reasons they might give for their action, and what they are likely to do in the future.

Social scripts govern people's mental states and behavior, just as stereotypes do, in ways that shape a person's self-conception and personal identity. In controlling mental states and behavior and shaping identity and sense of self, social scripts also decrease a person's autonomy.

How Internalizing Social Scripts Affects Identity

Oppressive social scripts are stories that are imposed upon a person that describe what kind of person they are, based on their possessing a morally salient and relevant trait, in a way that subjugates people who have that trait (Stoljar 2015, 115). When other people impose oppressive social scripts on a person, they make assumptions about what kind of person the person must be, given that they possess a certain trait, in a way that devalues them on account of that trait. These dominant scripts become a part of someone's identity that they have no control over, and may not consent to or choose, yet which nonetheless help shape and constitute who the person is. Even when a person wishes to repudiate the oppressive social script, they are compelled by its imposition to somehow react and respond to it. Ignoring it is not a choice because of the way it is forced upon a person. The force of the oppressive social script requires a reaction from the person as they are required to relate to it in some way, such as through acceptance or rejection.

Third-person conceptions of identity – conceptions of identity that are applied to a person by others – are often imposed on individuals irrespective of whether they agree with or endorse those conceptions. Regardless of what they themselves think of those perceptions of their identity, other people view the person's identity in terms of these third-person conceptions and thus impose them on individuals without the individuals' consent, choice, or control. It would be easy to say that a person who has a third-person conception of identity imposed on them by others that they do not wholeheartedly accept as belonging to them should be able to reject that conception and be rid of it. However, often these third-person conceptions of identity become intertwined with other aspects of a person's identity, sometimes aspects that they do choose or endorse, and it becomes impossible to separate them out.

For example, how a Black person in America is conceived of by others, with all the stereotypes and baggage that are associated with Blackness, becomes impossible to untangle from the self-concept a person forms about themselves. A person who views themselves as a Black American may want to reject the negative stereotypes that go along with that identity, but these can be confusingly intertwined with positive aspects of identity they want to retain, such as being part of a supportive community and being proud of their heritage. Moreover, what is viewed in the dominant culture as negative may be regarded in one's community as positive, making it difficult to tease out the positive and negative aspects of identity. For example, Black hair may be rejected as ugly in the dominant culture but welcomed as beautiful within the Black community; a person may develop conflicting feelings about their hair as a result.

In the context of mental illness, a person with mental illness may want to reject the negative stereotypes of mental illness that are imposed by others yet find them intertwined with their own uncertainty about how competent and dependent they are. Others' view of them as incompetent or dependent based on their mental illness may be intertwined with an uncertainty they have about their own experience. Or they may find that help-seeking is viewed as a sign of weakness within the dominant culture yet regarded as a sign of strength within the mental health community, making their own feelings about help-seeking confused. In one context a person may value the characteristic associated with them, while in another context they are taught that they should want to reject it. This can complicate their own self-concept.

In addition, even absent such intertwining, third-person conceptions of identity often become part of how a person sees themselves; this occurs even when they do not endorse these conceptions or believe they ought to apply to themselves. This is because part of how they see themselves is through the eyes of others, and they are compelled to react to whatever conceptions of identity others foist upon them. A person's self-concept is shaped in part by others' perception of them. When others see the person in a certain way, they can't help but see themselves in that way as well, even if they want to reject that conception. When they are forced to see themselves through that external lens, they find that they are compelled to react to that conception of themselves by accepting or repudiating it, proving or disproving it, and trying to live up to it or trying to show that they do not fit it. They are forced into a position where they have to respond to the third-person conception of themselves regardless of how they themselves view it and how they themselves view their own self-identity.

The relationship that they develop to the oppressive social script is thus more complicated than simple acceptance or rejection. A person often has no choice but to internalize the oppressive social script even if they do not endorse it; it becomes part of the person's psychological makeup even if they wish to reject it, and it calls on the person to react in various psychological ways. Stoljar says, "Oppressive social scripts are embedded in agents' psychologies in ways that call for multiple psychological responses – for instance, anticipation, adjustment, accommodation, and evaluation – that are not typically required from agents to whom neutral or 'dominant' scripts apply" (Stoljar 2015, 118). The imposition of oppressive social scripts calls for certain types of psychological reactions that other conceptions of who a person is do not call for.

If a dominant social script is imposed on a person, where they are seen as fitting social norms about who and what to be, the person can simply accept the script without further thought. Oftentimes, people do just that, which is why they can be so unaware of how they are seen as fitting dominant social norms and thus unaware of their power and privilege. They are not required to react to such scripts, so they are often unaware of the imposition of them. They just accept who they are as "normal" because they unwittingly fit social norms and give no further thought to it.

People who are subject to oppressive social scripts, on the other hand, do not have this luxury. Because they are forced to respond to the social script, they are compelled to give it thought, to be aware of its imposition on them, to be self-conscious and intentional about how they act in relation to them. This consciousness of the social script that is imposed on a person can make a person *self-conscious* and inspire the anxiety and stress that stereotype threat generates. Oppressive social scripts make a person aware that they can and will be judged according to the script and cause the psychological reactions that arise from being judged and from the uncertainty involved with how they are judged.

How Stereotypes and Oppressive Social Scripts Undermine Autonomy

In controlling people's mental states and behavior, stereotypes and oppressive social scripts cause many types of harm to people, including imposing constraints on people, limiting people's self-understanding and understanding of their situation, causing people to expend excessive amounts of mental energy and labor, making people more self-conscious, diminishing self-worth and self-esteem, inhibiting self-reflection and self-appraisal, and burdening people psychologically. Many of these harms can be understood as threats to autonomy. Here I explain how the power of stereotypes and oppressive social scripts to control people's thoughts, feelings, behavior, and social interactions diminish the autonomy of people who could be stereotyped in various ways.

Interference with Self-Determination

First, stereotypes and oppressive social scripts interfere with the self-determination aspect of autonomy by imposing psychological constraints on a person's ability to feel, think, and act the way they might want to.[2] Self-determination is the ability of a person to direct their life as they choose and the ability to make choices about what they value, what they want to do, and who they want to be. Marina Oshana describes self-determination as self-directedness, where a person can envision what kind of life they want to live and what kind of person they want to be, and set goals for themselves in relation to these ideals that they try to realize through their choices and actions (Oshana 2003, 100). Self-determination involves shaping the contours of one's life oneself so that a person can do

[2] Catriona Mackenzie identifies three aspects of autonomy: self-determination, self-governance, and self-authorization (Mackenzie 2014). I structure my discussion about autonomy in relation to these three dimensions.

what they wish and be whom they want to be based on their reasons and values.

In order for a person to have self-determination, freedom and opportunity conditions must be met. Freedom conditions include living within a social and political context that allows a person to have the political liberty to choose and act as they wish, as well as having the psychological freedom to choose and act according to their values and desires without external encumbrance that unduly shapes that choice (Mackenzie 2014, 17). Opportunity conditions include having sufficient opportunities available to a person in their social and political environment that enable them to be able to enact their choices and values and do what they wish to do, within reason (Mackenzie 2014, 17). Stereotypes and oppressive social scripts create a constraining social environment that prevents people from being able to direct their life as they choose by making it so that they are unable to enact what they choose to do and who they choose to be, and even unable to determine for themselves their desires and values about what they want to do and who they want to be. In other words, stereotypes and oppressive social scripts both prevent people from carrying out the actions that enable them to enact their choices and values and inhibit people from developing these choices and values in the first place. This violates both freedom and opportunity conditions of self-determination, making it so that individuals have insufficient freedom and opportunities to develop and exercise choice.

Stoljar identifies three ways that oppressive social scripts (and stereotypes) psychologically constrain people, and I add a couple other ways. One of Stoljar's observations is that stereotypes and oppressive social scripts require people who could be stereotyped to respond and adapt to them independently of people's relationship to them (Stoljar 2015, 118). The constant presence of cultural stereotypes forces people to acknowledge and respond to them regardless of whether people endorse them. Similarly, oppressive scripts compel people to adapt to the script even when they do not consent to, choose, or endorse the script. This kind of control constrains what choices a person can make within a given situation where stereotypes or oppressive social scripts govern interaction.

Moreover, stereotypes and oppressive social scripts compel a person to expend certain kinds of labor and shape what that effort looks like. Adaptation requires a level of motivation, energy, exertion, and self-control that must be expended, creating labor for the person that someone who is not subject to such scripts does not have. Stoljar notes that a person subject to an oppressive social script has to form plans and intentions based on this script and that it occupies a good portion of the person's mental life (Stoljar 2015, 118). Even when a person is not conscious of how the script applies to them or is imposed upon them, it takes up mental space unconsciously through implicit bias. Adaptation often involves working hard to try to prove that the oppressive social script does not apply to the person and often results in either overwork or disengagement, just as reactions to stereotype threat do. The amount of mental space that stereotypes and oppressive social scripts occupy within a person's mind makes the person focus on the aspects of their experience related to these and ignore other aspects of their experience. This imposes severe constraints on what a person can choose to do and be.

Another way that stereotypes and oppressive social scripts impose psychological constraints on a person is the way they compel a person to respond to the evaluation of them that is invoked by the script or stereotype (Stoljar 2015, 118). Stereotypes and oppressive social scripts involve an automatic evaluation of the person upon whom they are imposed. When the person appears to fit a negative stereotype that is foisted upon them, either on its own or embedded within a larger script, they are judged as bad because the stereotype is seen as bad. When the person appears not to fit the stereotype associated with the script, they are judged as an exception to the badness. Possibly they may be seen as good, because it is good to not fit the stereotype, and it may be praiseworthy to be seen as overcoming a negative attribute, but sometimes the judgment does not rise to this level as it may involve simply seeing the person as escaping being bad. Both stereotypes and oppressive social scripts thus automatically involve taking an evaluative stance toward the people upon whom they are imposed. The person is always and inescapably being judged by others in light of the script or stereotype.

As part of the necessarily evaluative aspect of stereotypes and oppressive social scripts, a person must take an evaluative stance toward themselves as well, assessing whether they think the stereotype or script applies to them and deciding how to respond to it whether it does or not. In other words, the evaluative stance thus is not simply taken up and imposed upon the person by others but also must be taken up and imposed upon oneself by oneself. The person subjected to stereotypes and oppressive social scripts is forced to make a self-appraisal, judging themselves in relation to the stereotype or script, and to respond to that judgment by taking some form of action. They have to choose whether to engage in behavior that seems to fit the script or whether to act in ways that seem to defy it. If they defy it, they must develop justifications for not adhering to the script or stereotype because the evaluation embedded within it will make them answerable to others. Thus they have to expend mental energy thinking about how and why they should act in certain ways, not to benefit themselves but rather to benefit others who expect them to act a certain way as part of how the script or stereotype is imposed upon them.

Stereotypes and oppressive social scripts are unavoidable when they exist within a culture. A person might have reasons for acting that have nothing to do with the script or stereotype, yet the script or stereotype nonetheless plays a role in shaping their choices and behavior. They are forced to respond to the evaluation embedded within the script or stereotype, both the evaluation others make and their own evaluation, even if they want nothing to do with it. It becomes part of their reasons for acting regardless of whether they consent to its role in their mental life and behavior.

An additional way that stereotypes and oppressive social scripts impose psychological constraints on people is that they require that individuals respond to the script or stereotype in order to attain equal access to what Elizabeth Anderson calls "cooperative interactions" or what we can understand as social contexts where people engage in moral and epistemic practices with each other (Anderson 2010, 53; Stoljar 2015, 118–119). To be treated as a moral and epistemic equal in moral, epistemic, and social interactions, a person has to not only disavow the script or stereotype and reject it as an accurate way to view themselves but they also have to act in ways that counteract what the script says or what the stereotype invokes. If the script or stereotype says that Black men are violent, or people with mental illness are violent, a person subjected to the stereotype or script has to prove that they are not violent through intentional choices about how to act in ways that make White people or people in the community feel safe so that they can be treated as moral and epistemic equals. If they do not take this intentional action, they will be viewed as violent even if they do nothing violent, simply because they did not take intentional action to disprove the script. They will be assumed to fit the script or stereotype unless they prove otherwise. And proving otherwise is the ticket to being regarded as a moral and epistemic equal who can participate with authority and credibility in moral and epistemic practices with others.

Not making the effort to prove that one does not fit the oppressive social script or negative stereotype will cause those who impose the script on the person to judge them as fitting the script or stereotype, because the default position that those in power tend to take is that people who have a certain trait automatically fit the scripts and stereotypes related to that trait unless proven otherwise. A person subject to oppressive social scripts or negative stereotypes is thus always burdened with the responsibility to decide how to respond to it and to expend energy and effort in trying to disprove that the script or stereotype applies. As already noted, disproving the script or stereotype does not necessarily result in a judgment that the person is good, and therefore a moral and epistemic equal, only that the person is an exception to the general rule that people like them are bad. However, being seen as an exception can be sufficient to be treated as a moral and epistemic equal. And putting in the work to disprove that one fits the script or stereotype is necessary for being regarded as an equal who can participate with authority and credibility in moral and epistemic practices, and social interactions in general.

Stoljar does not discuss the way that oppressive social scripts make people *aware* of having the script imposed upon them in a way that people who are subject to dominant scripts are not, and that this awareness causes self-consciousness, anxiety, and stress. However, this constitutes an additional way that stereotypes and oppressive social scripts impose psychological constraints on a person as well as impair the competence and authenticity conditions of self-governance, which I discuss in the section titled Threats to Self-Governance. When a person becomes aware that a stereotype or social script is being imposed on them, they become self-conscious and self-aware about what the script or stereotype says about them and how they will react in response. This self-awareness occupies significant mental space and demands a considerable amount of self-control.

Since a script ascribes motivations to a person's behavior, predicts their behavior, and offers interpretations of their behavior based on presumed causes, the person who becomes aware of a script has to think about how their behavior will be interpreted and exercise significant self-control to ensure that their behavior is expressed in a certain way so that it can be interpreted as the person desires. They have to be very careful in what they say and how they say it so that they can have some control over how their speech is heard and interpreted. An ordinary amount of self-control will often not suffice to control one's speech and behavior enough for it to be interpreted as one wishes, usually to counter the oppressive social script. Rather, an extraordinary amount of self-control is required, more than what many people have.

People who have mental illness, for example, sometimes do not have a whole lot of control over their behavior. Many people with mental illness do not even have enough insight into their condition to recognize that their behavior and speech might be interpreted in certain ways, so it would not even occur to them to exercise self-control over these for the sake of influencing the interpretation. Even when people do have insight, recognize that they are ill and that others see them as well, and see that an oppressive social script is being imposed on them, garnering the amount of control necessary to regulate their behavior and speech may be extremely difficult. This is due in part to the strong influence of mental illness symptoms on behavior that a person may not be able to counter. Some people simply cannot regulate themselves sufficiently. For these people, the oppressive social script will be imposed on them regardless of what they do; they are unable to prove to others that they can be seen as an exception to the script.

When people try to refute a script by acting contrary to its predictions and expectations, the excessive effort required to exert self-control may be too much for a person, and they may give up. When they give up, they are essentially letting the script shape how people interpret their behavior because it seems too difficult to challenge it. Giving up is easier. When someone gives up, they may respond to the difficulty inherent in challenging the script by willfully enacting the script, behaving as they are predicted and expected to. They may think something along the lines of, "Oh, what the heck; since I can't fight it, I might as well live it out." They may even exaggerate the behavior that is predicted by the script so that they appear as if they *fully* embody the embedded stereotype.

Willfully enacting the script is a way for a person to exert power in a situation where they have very little power. They do not have the power to control their behavior in a way that challenges the script, so they wield what little power they have to do what they *can* do, to fully embody the stereotype in the script. In this way, they can feel like they are exercising some agency rather than having their agency fully scripted by other people in a way that is out of their control.

To the extent that a person can control their behavior and speech, however, they will often try to do so in a way that challenges a script or stereotype when they become aware of an oppressive social script or a negative stereotype being imposed on them. When people are aware that oppressive social scripts or negative stereotypes are being foisted upon them, they often do not *want* the script or stereotype to be inflicted upon them; they do not want their behavior to be judged in accordance with it. Thus they try to resist to the extent that it is possible for them to do so. They try to act in ways that counter the script or stereotype. They have to think through what actions would demonstrate a different interpretation of their motives so that they can be seen in a different light. They have to consider what actions would disprove the stereotype embedded within a script, considering all the ways an action could be interpreted in given contexts. This thought process of determining how to resist a script or stereotype effectively takes up considerable time and energy. Then they have to actually exert whatever control over their speech and behavior that they can in order to try to get other people to interpret their behavior in the way that they would like it to be interpreted.

Not only do they have to exert control over their speech and behavior, but they also have to become conscious of and think through their reasons for acting so that they can be in a position where they can share these. They have to acquire a self-awareness of their reasons for acting that people often do not have; people often act intuitively, unconscious of their motives and reasons, but they do not have the luxury of being able to do this if they can help it. They must be ready to share those reasons in a calm, rational, unemotional, and controlled way that will hopefully be heard and taken seriously by others. They want to be seen as agents capable of giving reasons for their action so that their agency and normative authority is respected by others and they are afforded credibility.

The immense self-awareness and self-control that oppressive social scripts and negative stereotypes create for a person is another way that scripts and stereotypes inhibit self-determination by preventing freedom conditions of self-determination from being met. By taking up so much mental space; making a person pay strict attention to their thoughts, feelings, and behavior; compelling a person to figure out how to respond; and requiring a person to expend considerable energy and effort through attempts at self-control, negative stereotypes and oppressive social scripts prevent a person from being able to enact choices and values they would otherwise pursue, and even prevent a person from being able to develop their choices and values. Instead, these stereotypes and scripts impose options, values, and social norms onto a person that shape how the person feels, thinks, and acts, limiting their freedom.

In requiring a person to respond to the script or stereotype that is foisted on them, oppressive social scripts and negative stereotypes act as external factors that unduly limit what choices for action a person can take, violating freedom conditions of self-determination. They circumscribe options so that a person is given the option of embodying the script or stereotype or challenging it in certain ways, but no option to ignore it and act however one wishes. The script or stereotype always compels a person to feel, think, and act in certain ways. Thus it exerts psychological force upon a person, constricting their psychological freedom in such a way as to violate the self-determination aspect of autonomy.

The fear involved with stereotype threat, of wanting to avoid being judged according to negative stereotypes and working hard to try to defy negative stereotypes (sometimes so hard as to decrease competence, which can paradoxically make a person appear to fit a stereotype), also restricts what options a person believes are available to them. This violates opportunity conditions of self-determination. When a person is motivated to act purely out of fear, they believe that their only options are to counter that fear by resisting what they are afraid of, or to give in to that fear by succumbing to what they are afraid of. A person who fears being judged according to a negative stereotype believes that their only options are to work harder so as not to appear to confirm the stereotype or to give up trying and succumb to the stereotype. Either way, the person's only apparent options are to react to the existence of the stereotype. They do not seem to have other options that do not involve reacting to the stereotype. This unduly limits what they can do, creating external constraints on how they can direct their lives. Thus, stereotype threat also violates opportunity conditions of self-determination, limiting what options are available to a person so they have fewer opportunities to act in different ways as they might choose had they had more options.

Threats to Self-Governance

Negative stereotypes and oppressive social scripts also inhibit the competence and authenticity conditions of the self-governance aspect of autonomy. Self-governance involves making choices that are one's own, rooted in a person's own reasons and values. Marina Oshana describes self-governance not only as having values, preferences, and commitments but also as "directing one's life according to such values and preferences" (Oshana 2003, 101) or being able to act successfully in carrying out one's values and commitments. In other words, it involves having a certain kind of power, "the power to determine how [one] will live" (Oshana 2003, 101).

In order for a person to have self-governance, competence and authenticity conditions must be met (Mackenzie 2014, 17). Competency conditions of self-governance are rational, cognitive, affective, volitional, and valuational capabilities that enable a person to develop and draw upon reasons and values that are their own, that they self-authorize as coming from them and not from someone or somewhere else. Authenticity conditions include the capacity for self-reflection and self-appraisal involved with reflecting on one's choices, reasons, and values and endorsing them as one's own. Another way to understand authenticity in this context is in terms of voluntariness of choice (Feinberg 1986, 104; Feinberg 1983, 7), where a choice is considered voluntary if it can be reflectively endorsed (Dworkin 1988, 20; Frankfurt 2003; Watson 2003). Negative stereotypes and oppressive social scripts interfere with the competence and authenticity conditions of self-governance by decreasing people's rational, cognitive, affective, volitional, and valuational capabilities and interfering with their ability to self-reflect about their choices.

As discussed in relation to stereotype threat, when a person becomes aware of the imposition of negative stereotypes or oppressive social scripts upon them, the resulting self-consciousness, anxiety, and stress can impair a person's rational, cognitive, emotional, volitional, and valuational functions, causing decreases of competence. If a person becomes less competent, more dependent, or more abnormal as a result of the psychological processes involved with being aware that they could be judged incompetent, dependent, or abnormal, then they lose some of the competence capacities involved with making an autonomous choice. Their rational capacity may become impaired, their ability to form and revise beliefs in relation to reasons and evidence may be challenged, their emotions may become dysregulated, they may be unable to reflect upon their desires and values, and their self-control can become defective. All of these capacities can become impaired when a person has an excessive amount of self-consciousness, anxiety, and stress, as well as when they are trying to compensate for appearing incompetent by working too hard. In these ways, stereotype threat – being aware that others may judge a person for fitting a negative stereotype – can violate competence conditions of self-governance.

Stereotype threat and the threat of having oppressive social scripts applied to oneself also violate authenticity conditions of self-governance by thwarting self-understanding, self-reflection, and self-appraisal. Impairments in mental functioning caused by trying to defy a stereotype, the stress of being preoccupied with figuring out how to respond to a stereotype or social script, and the mental space that this deliberation occupies all inhibit self-understanding, self-reflection, and self-appraisal. Insufficient self-understanding, self-reflection, and self-appraisal make it so that a person cannot reflect on their desires, values, and choices and decide whether to endorse them or not. This makes the authenticity condition of self-governance, where one's choices, values, and desires are truly one's own, unable to be met.

Paul Benson explains how stereotype threat disrupts authenticity conditions (Benson 2015). Benson understands authenticity in terms of John Christman's account, where a desire or choice is authentically one's own when one would endorse the development of the desire or choice upon reflection if one were motivated to reflect on it (Christman 2004, 154). Benson writes: "Persons are autonomous in their choices and actions, according to Christman, when they act on motives whose processes of development they would not resist or disown upon suitable reflection" (Benson 2015, 129).

This requires the ability to imagine oneself in different situations with different reasons that could govern one's actions. Thus, Christman says, "a person reflects adequately if she is able to realistically imagine choosing otherwise were she in a position to value sincerely that alternative position" (Christman 2004, 154),[3] which can only be done when one is free from social and psychological conditions that inhibit self-appraisal. If a person is oppressed, this oppression prevents them from self-appraising appropriately and prevents them from being able to imagine-making a different choice in a different situation, based on different reasons. Oppression leads people to feel as if they have no choice, that they would act the same way in any circumstance, for the same reason.

A person who fears that they will confirm a negative stereotype or oppressive social script is oppressed by that stereotype or script because its existence shapes the person's action on account of fear of being judged. Even if the person does not believe that the relevant stereotype or script is true, or that it applies to them, knowing that other people may judge them according to that stereotype or script influences what actions a person takes and what they devote their time, energy, and resources toward, and consequently what psychological processes they will be subject to. They will take action that involves trying not to uphold the stereotype or script regardless of their actual circumstances and regardless of whether actual individuals would actually judge them. It is the fear of being judged, based on the potential that one could be judged, that drives the action, not an actual judgment that someone makes. A person motivated by fear of others' judgment is too restricted by that fear to be able to imagine a situation where they would *not* be judged by others, and too restricted to be able to choose willingly to engage in actions or appear in ways that

[3] In later work, Christman amends this criterion of authenticity in terms of not being alienated from one's autonomously chosen values and desires, and the history and processes by which they were formed, upon critical reflection (Christman 2009, 142–149).

would seem to confirm the stereotype one fears. The stereotype creates psychological conditions that make it impossible for a person to imagine making an alternative choice based on different reasons.

If a person subject to stereotype threat were able to reflect upon their choice of how to react to the potential of being judged, free from the psychological constraints that restrict how they understand the nature of their choice, they would not endorse the development of that choice. They would not choose to try harder to resist a stereotype or script if the stereotype or script did not have undue pull over them. They would not choose to lose confidence with themselves or disengage from the activity if the stereotype or script carried no weight. Thus, their reactions to the potential of being judged by the stereotype or script are wholly informed by the existence of that stereotype or script and by no other intrinsically motivating reasons. Their reactions to the stereotype threat to which they are subject would be unendorsed upon reflection and thus are not authentically their own. In this way, stereotype threat and the threat of having oppressive social scripts foisted upon a person violate authenticity conditions of self-governance.

Another way that oppressive social scripts in particular violate competence and authenticity conditions of self-governance is by limiting a person's self-understanding and understanding of their situation. As discussed in Chapter 1, Śerife Tekin notes that the social scripts that emerge from what she calls "DSM culture" – where mental disorders and mental illness experience are understood primarily or solely in terms of DSM (*Diagnostic and Statistical Manual*) criteria – delineate how a person views their own experience (Tekin 2014; Tekin 2011). Typically, this involves focusing on symptoms, as well as biological and psychological factors that affect mental states and behavior, and ignoring interpersonal and relational aspects of experience. (Tekin does not mention other aspects of experience that also get ignored, including existential and philosophical aspects, religious and spiritual aspects, and socioeconomic and political aspects. These, too, can be important for a person's self-understanding and understanding of their situation, but are completely overlooked when a person adopts a primarily DSM view of their experience.) The result is a loss of insight and a loss of self-understanding, where a person's understanding of their experience gets flattened and reduced to DSM criteria and frameworks instead of being enriched by an appreciation of the multitude of factors, including social and relational factors, that affect experience. Lessened self-understanding and understanding of one's situation reduces a person's rational, cognitive, and valuational abilities, and possibly their affective and volitional abilities, consequently diminishing their competence as well, and thus their capacity for self-governance.

Insufficient self-understanding and insight in turn impede self-reflection and self-appraisal, which violates authenticity conditions of self-governance. Lowered self-understanding and understanding of one's situation interferes with processes of self-reflection and self-appraisal by making a person unable to see themselves and their situation in all its nuance and complexity. Without a properly comprehensive understanding of themselves and their situation, they are unable to reflect upon their choices, desires, and values sufficiently to be able to figure out whether they endorse these as their own. They also cannot evaluate themselves in light of their choices, desires, and values because their view of themselves is too narrow and reductive. Inadequate self-reflection and self-appraisal hinder authenticity, and thus self-governance.

Inhibition of Self-Authorization

Negative stereotypes and oppressive social scripts also intervene in the normative authority and epistemic credibility aspects of self-authorization of one's choices. Self-authorization consists of having the normative and epistemic authority to be and to be seen as self-determining and self-governing (Mackenzie 2014, 18). When a person has self-authorization, they are regarded as an authority over their own life. This includes having and being seen as having authority over their choices, values, desires, thoughts, feelings, and actions.

Marina Oshana says that to be autonomous in the self-authorization sense "is to have a kind of authority over oneself as well as the power to act on that authority" (Oshana 2007, 411). Having this authority requires having a certain self-conception: a conception of oneself *as* an authority, seeing oneself as having the normative authority and epistemic credibility to be able to speak for oneself and be an author of one's actions (Mackenzie 2008, 514). Catriona Mackenzie (2014) and Natalie Stoljar (2015; 2011) only talk about normative authority, but I would argue that epistemic credibility is just as important for self-authorization as normative authority is (see Roessler 2015). Natalie Stoljar goes further than this to note that normative authority (and epistemic credibility) are intersubjective concepts. Being recognized *by others* as having this authority (and credibility) to speak for oneself and author one's choices and actions is conducive to actually having this normative authority (and epistemic credibility) (Stoljar 2011, 376).

Some of the conditions required to exercise self-authorization and to both see oneself and be recognized by others as having authority over one's own life include accountability; self-evaluative attitudes such as self-worth, self-respect, self-trust, and self-confidence; and social recognition (Mackenzie 2014, 19). When these conditions are met, a person can both see themselves as having and be recognized by others as having normative authority over their choices, values, desires, and actions; and they can both see themselves as having and be recognized by others as having epistemic credibility in being able to speak for themselves. Negative stereotypes and oppressive social scripts interfere with these conditions in many ways.

When a person is subject to negative stereotypes, including those embedded within oppressive social scripts, this causes other people to not recognize their epistemic credibility. They are seen as having less credibility in being able to speak for themselves, and their views, as well as their reasons for action, are often ignored, dismissed, and discounted. Thus, the person subject to stereotypes might be capable of giving reasons for their action that they have reflected upon and that are their own, and not necessarily or simply the motivations that the script ascribes to them, but these reasons are not taken up by others. When a person's reasons for acting are not listened to and taken seriously by others, the person's epistemic credibility is denied, and they are dismissed and discounted. When this discounting occurs due to the negative stereotypes embedded within the oppressive social script, it constitutes epistemic injustice based on identity-prejudicial credibility deficit (Fricker 2007, 17–29).

The testimonial injustice of not having one's reasons for acting taken seriously by others due to the person having mental illness is one form of the psycho-pathophobic epistemic injustices discussed in Chapter 2. It occurs in transactional, institutional, and structural contexts where people with mental illness interact with other individuals, interface with institutions, or have their actions enabled and constrained by social structures. Psycho-pathophobia shapes personal and institutional interactions through attitudes such as aversion, banality, callousness, insensitivity, and untruthfulness, all of which treat people with mental illness poorly and deny them the credibility of being able to speak for themselves, give reasons for their action, and interpret their experience (Kidd 2019; Kidd and Carel 2021, 74). Psycho-pathophobia also influences the way actions get channeled through social structures that prevent people from acting and speaking for themselves in some ways and coerce them into acting and speaking in other ways.

When a person is unable to give reasons for their choices and actions that are taken seriously by others, they lose normative authority. Andrea Westlund notes that having normative authority and being seen as having normative authority involve in part an intersubjective component of being answerable for one's action-guiding commitments, or the values that underlie one's choice (Westlund 2009, 33). To have normative authority, a person must be responsive to the reasons given by others and must be able to answer others' questions about their own reasons for action, almost as if they can give a defense of their reasons. Through dialogue, a person must be capable of giving reasons for their choices and action, listening to and responding adequately to reasons given by others, and answering others' inquiries about their own reasons in a way that makes sense to others. Being answerable in this way requires that others take the person seriously, listen to their reasons, and engage with them about those reasons. If a person is not afforded the credibility involved with taking them seriously, they will lose out on this opportunity and be unable to exercise the intersubjective part of normative authority.

When a person's reasons for acting are not listened to and taken seriously by others, their authority to decide for themselves how and why to act is not recognized and respected by others. This diminishes their normative authority. If they have a positive self-conception and a strong sense of self-worth and self-respect, they may still feel like they have normative authority to authorize their own choices even though it is not recognized by others. This can be sufficient for them to act as agents capable of giving reasons for their action. However, their normative authority remains diminished when it is not recognized and respected by others. Negative stereotypes and oppressive social scripts that deny the epistemic credibility and normative authority of individuals upon whom these stereotypes or scripts are imposed decrease the individual's ability to self-authorize their choices as their own.

Conclusion

Stereotypes and oppressive social scripts undermine autonomy in various ways because of the way they have the power to control people by controlling their thoughts, feelings, behavior, and social interactions. This power to control people is an expressive harm because it exists wherever cultural stereotypes exist, independently of whether any particular individuals endorse the stereotypes. Cultural stereotypes permeate the atmosphere so they always govern social interaction, even if the people involved in the interaction dispute them. In being ever present, individuals often have no choice but to internalize cultural stereotypes as they always feel the potential for being judged according to these stereotypes. Even in a setting where no individuals endorse or consciously act on negative stereotypes, a person who could be subject to a stereotype can always anticipate the *possibility* of being judged, and this is sufficient for the stereotype to affect their thoughts, feelings, behavior, and interaction.

The expressive harm of stereotypes helps us understand why people with mental illness feel like mental illness stigma is pervasive even when no actual acts of stigmatization occur, and even in environments where individuals repudiate negative stereotypes about mental illness. The ever-present nature of cultural stereotypes makes it so that people with mental illness can anticipate the possibility of being judged according to negative stereotypes and oppressive social scripts, and thus feel the existence of stigma, even when others do not actually hold or express that stigma.

Negative stereotypes and oppressive social scripts violate several conditions of autonomy. These include the freedom and opportunity conditions necessary for self-determination, the competence and authenticity conditions required for self-governance, and the epistemic credibility and normative authority conditions of self-authorization. Because of their role in undermining autonomy, negative stereotypes and oppressive social scripts regarding mental illness must be challenged, resisted, and changed so that people with mental illness no longer feel the weight of their being always in the atmosphere. To enhance the autonomy of people with mental illness, it is paramount that society addresses the cultural pervasiveness of mental illness stigma.

Because of the persistence of mental illness stigma, many people who have mental illness learn to internalize stigma by either projecting it onto themselves through self-stigma or by always being aware of the possibility of being judged by others through attendant stigma. In the next chapter, I examine how internalized stigma of both kinds can lead people with mental illness to internalize the social norms embedded within negative stereotypes and oppressive social scripts and consequently adopt adaptive preferences that are harmful to their well-being, irrational, unreflective, and nonautonomous.

CHAPTER 4

Internalized Stigma and Adaptive Preferences

Introduction

Sometimes stigma causes people to develop harmful adaptive preferences that inhibit their ability to grow, recover from mental illness, and live a meaningful life by trapping them in an overly restrictive life pattern. Mental illness stigma can give rise to adaptive preferences when people internalize the stigma so that they endorse and accept the social norms associated with the stigma, leading people to feel obligated or resigned to try to live up to those norms. When these social norms are harmful, feeling bound to live up to them can imprison a person within a life that is too restrictive, that inhibits their flourishing, and that they wouldn't really want if they felt like they had other options.

In Chapter 1, I explained that when people internalize stigma, they either endorse the stigma as applying to themselves (what is called "self-stigma") or they see how others may apply the stigma to themselves and accept this application as inevitable (what I called "attendant stigma"). In either case, they are aware that they could be seen as fitting negative stereotypes and aware of themselves as possibly apt candidates for prejudice and discrimination. With self-stigma, they form their own judgment that they conform to negative stereotypes; with attendant stigma, they are aware that others may form this judgment. This awareness pervades their consciousness, so they are self-conscious and often vigilant about how they act and how they come across to others. When they see that negative stereotypes can be applied to them, they might incorporate aspects of these stereotypes into their identity, as I explained in Chapter 3. Incorporating aspects of stereotypes and prejudice into one's identity, and endorsing stereotypes as true and applying to oneself, are forms of internalizing stigma.

Internalizing stigma often involves internalizing social norms that are associated with the stereotypes that underlie stigma. This internalization occurs through the incorporation of these norms into a person's understanding of their experience, motivations in acting, and self-concept and personal identity. This chapter begins by examining the effect of internalizing social norms embedded within negative stereotypes on well-being and autonomy. Then I analyze the desires and values a person forms accordingly in terms of adaptive preferences, showing that such preferences are harmful because they often hinder well-being and flourishing, tend to be oppressive, and may be irrational and non-autonomous. Let us look at the relationship between negative stereotypes and harmful social norms.

Stereotypes and Social Norms

Negative stereotypes are action-guiding in both a backward-looking way and a forward-looking way in that they shape how a person understands actions they have done in the past and direct how they act in the future. Embedded within stereotypes are social norms that prescribe what kind of person they should be and how they should live their life. When a person is seen as fitting a negative stereotype, their past behavior gets interpreted in the light of this stereotype, so that their action is summed up in terms of what kind of person the stereotype indicates that they are. Stereotypes also direct what action a person takes in the future by setting up expectations for how they are likely to act based on what kind of person the stereotype says they are. When a person internalizes a negative stereotype, they learn to see their past action in light of the stereotype and come to see themselves as a certain type of person who tends to act in a manner conforming to the stereotype. In this way, a person who internalizes a negative stereotype understands their experience, behavior, and identity as framed by the stereotype and directs their action in accordance with the stereotype.

Let us examine how this works with self-stigma. Self-stigma involves endorsing negative stereotypes as true and forming a judgment that negative stereotypes apply to oneself. When a person endorses negative stereotypes, they accept the social norms that are embedded within the stereotypes as norms that should guide people's action and norms that can be used to evaluate past behavior. When a person applies negative stereotypes to themselves that they endorse as true, they accept that the social norms embedded within the stereotypes should guide their own action and accept that these norms serve as a basis for evaluating their own behavior.

For example, consider the negative stereotype that people with mental illness are incompetent.[1] A person with mental illness who internalizes the stigma associated with this stereotype believes that the stereotype is true – people with mental illness *are* incompetent – and that the stereotype applies to themselves. Thus, they believe that *they* are incompetent on account of having mental illness. This guides how they understand themselves and their experience, making them interpret their experience and past actions in relation to the belief that they are incompetent, seeing, for example, certain actions or experiences as evidence of incompetence. This belief also guides how they will act, for example refraining from certain actions that they believe they are too incompetent to execute successfully or lowering their expectations for themselves and restricting their actions accordingly.

The belief that they fit a negative stereotype can arise in many ways. The evidence may support the belief; their experience of their mental illness may be that it makes it so difficult to function that this seems to support the idea that they are incompetent. In this case, their endorsement of the stereotype is consistent with the evidence, which is their experience. They may believe the stereotype *on account of* the evidence, or they may believe the stereotype independently of the evidence, depending on what role the evidence plays in their belief formation, while the evidence nonetheless appears to confirm their belief. For people whose belief is supported by the evidence of their experience, it can be especially difficult to challenge such a belief because it can be difficult if not impossible to argue against what they take to be evidence.

Contrarily, the evidence may undermine the belief; their experience of their mental illness may be that it is difficult, but that they can function nonetheless, which would refute the idea that they are incompetent. However, people do not always pay attention to evidence that they should, especially when their epistemic capacities are compromised. A person may find themselves being able to function in various areas of their life yet not be able to see that this counts as evidence that they are not incompetent. They may believe that they are incompetent despite the available evidence.

[1] As noted in the Introduction and Chapter 1, competence here refers to a person's ability to function in various life domains such as social interactions, social relationships, work, school, civic life, and leisure activities. Craig Edwards (2010) defines competence as being able to perform a task effectively; I interpret this more broadly as being able to function in specific areas. This is different from the narrower understanding of competence as decision-making capacity and the broader conception of competence used in the stereotype content model as encompassing capability, assertiveness, agency, intelligence, skill, efficacy, independence, confidence, achievement, prestige, and competitiveness.

This kind of self-delusion, of sustaining a belief despite its being refuted by available evidence, is not uncommon. If they are not able to recognize evidence that refutes the belief, or if they are unable to interpret the evidence properly, they may hold the belief regardless.

In any case, the belief that they are themselves incompetent, based on the stereotype that people with mental illness are in general incompetent, entails certain social norms that are embedded within the stereotype. These social norms include the value of competence and functioning, the importance of overcoming difficulty, the shame of dependence on others for participating in daily life activities, and the shame of failure. Some of these norms are ideals that one should try to realize, while others are norms that describe what one should try to avoid or refrain from. These norms set up expectations for how a person should act. In the case of norms surrounding the stereotype that people with mental illness are incompetent, it is presumed that people should be competent and be able to function in various life domains; they should be able to overcome obstacles; they should be able to participate in activities with relative independence; and they should be able to act successfully in creating the outcomes they strive for.

When a person is judged as to whether they conform to the stereotype of being incompetent, they are judged as a good or bad person based on whether they live up to the social norms embedded within the stereotype and avoid the sources of shame and failure that are associated with the norms. When a person judges that the stereotype of mentally ill people being incompetent fits themselves, they internalize these social norms as norms they should follow, that should guide their action. They use this to evaluate their past actions and evaluate themselves as a good or bad person correspondingly, as well as to direct what they do in the future. These norms thus serve as a basis for self-evaluation as a good or bad person and as a basis for creating expectations for how they will act.

Social norms are action-guiding in the case of attendant stigma as well. When a person does not endorse the stereotype of mentally ill people being incompetent as true, they may nonetheless recognize how others may see it as true and see how it could be applied to themselves. The norms underlying the stereotype serve as a basis for *others'* evaluation of the person as good or bad and as a basis for *others'* expectations of how the person will act. When a person recognizes how the stereotype will be used for evaluation of and for setting up expectations for the person, they may come to accept or at least be resigned to the fact that the stereotype will be used in these ways. This will force them to pay attention to how the stereotype and its embedded social norms structure their mental states, behavior, and social interaction. This acceptance or resignation is one of the ways that stigma gets internalized as attendant stigma, when a person is aware of and pays attention to stigma and stereotypes that thus shape who they are and what they think, feel, and do.

Negative stereotypes play both a backward-looking role and a forward-looking role, both explaining behavior that a person has done and creating expectations around what a person will do. For example, in relation to explaining behavior, a person who believes that they are incompetent will come to interpret their experience with difficulties with functioning in various life domains as evidence of incompetence, rather than seeing it as an adversity that they can learn or grow from, as a momentary setback that can be overcome and moved on from, or as a normal difficulty that nearly everyone faces at some point in their lives. Their difficulty with obtaining a job will be interpreted as proof of incompetence rather than an effect of constraining economic circumstances that they have no control over. When other people believe that a mentally ill person is incompetent, they will interpret the mentally ill person's experiences with difficulties with functioning, or difficulty obtaining a job, as evidence of incompetence in the same way, precluding other possible interpretations of that experience.

The negative stereotype that a person with mental illness is incompetent also sets up expectations that the person with mental illness will not be able to function in important areas of their life. When a person believes themselves to be incompetent, they will form expectations for themselves that they will not be able to overcome obstacles and that they will fail in whatever they try doing. When others believe that the person with mental illness is incompetent, these others will form similar expectations, which then get adopted by the person themselves when they internalize the stigma that is imposed upon them. These expectations are action-guiding in that the person with mental illness will form desires and values around the expectation and make choices and take action with this expectation in mind.

How Internalizing Negative Stereotypes Restricts Options

When a person internalizes the belief that people with mental illness are incompetent, and that this means *they* must be incompetent, too, or at least that they are seen by others as incompetent, this expectation takes certain options off their deliberative screen and puts others on it. Because of this belief, the person may assume that they cannot work at certain kinds of jobs, have certain types of relationships (such as a spouse or children), or participate in certain activities. They will limit their desires to what they believe is possible, so they will abandon desires that they may have had that now seem impossible, removing these options from their deliberative screen.

They may adapt to their limited options and desires by changing their value set so they no longer value what they feel they cannot attain but instead limit their values to what seems attainable. For example, if having a life partner or children seems impossible, they may stop valuing having a life partner or children and instead value something that seems more attainable, such as having a pet. As their values change, so does the way they direct their life. They no longer make plans that would support having a life partner, such as joining clubs to try to meet new people; or having children, such as child-proofing their living space or getting a flexible job that would enable them to do caretaking duties. They might make plans that would support their new values, such as buying a pet enclosure or preparing to wake up early to feed their pet.

Sometimes when a person's values change, their new values can be just as positive as their former values. Having a pet is a good thing for many people, as it can contribute to flourishing and living a meaningful life, just as having a life partner or a child can be. However, sometimes a person's new values, or at least how they conceptualize those new values, can be harmful to them. If a person gives up the value of having a life partner or children, they may replace that value with a value of being single, but then interpret a single life as a lonely life without social relationships. It is not that they *want* to live a lonely life, but they may feel as if they have no other options, and they may become resigned to such a life. This resignation can lead to social withdrawal and isolation, creating the lonely life without social relationships that they believe they are destined for. In this way, a person's change in values can be harmful to them.

The more the incompetent a person expects they will be, the more the options appear unattainable, and the more the desires and values will change to accommodate this lack of options. A person who believes that they cannot successfully create the outcomes they desire through their own action will self-select to reduce the range of what outcomes they can desire and value. They will lower their expectations of what is possible, and they will shift their desires and values to accommodate this more limited range of possibilities. Stereotypes that have embedded within them social norms that restrict what a person feels capable of doing guide action in very restrictive ways in how they structure values and desires. When a person's values and desires get downgraded and narrowed in this way, this decreases the range of what action a person feels capable of, removing certain options that seem impossible off the deliberative screen, and reducing the range of options that a person can act on. In this way, negative stereotypes decrease agency.

By restricting options, the internalization of negative stereotypes also hinders autonomy. As discussed in Chapter 3, one set of conditions for the self-determination aspect of autonomy are freedom conditions in which certain personal and political liberties exist that allow a person to have the freedom to make the choices they wish to make (Mackenzie 2014, 17). Oppressive social norms violate freedom conditions of self-determination by reducing options through the coercive shaping of desires. Social norms shape a person's desires and values, including their conception of the good and what kind of life they want to live (Ells, Hunt, and Chamber-Evans 2011, 86). Oppressive social norms coerce choices by shaping desires in unjust ways, removing some options for what can be desired off the deliberative screen and inserting others that unjustly restrict how a person can direct their life and what kind of life they wish to live.[2] A person's desires are typically constrained to what seems possible as people adapt to limiting circumstances by directing their desires in accord with those limitations (Stoljar 2014, 233–234).

In addition to the negative stereotype that mental illness makes a person incompetent, other negative stereotypes include the view that people with mental illness are needy and can't take care of themselves, that mental illness is a long-lasting or permanent condition, and that mental illness gets progressively worse over time. Internalizing these stereotypes can lead a person to believe that many options are not available to them, such as acting on their own behalf (if they believe they are overly dependent on others), or the possibility of managing their illness so it does not rule their life (if they believe their illness is permanent). These stereotypes can also make a person feel hopeless about their future, eroding their sense of agency by undermining their belief in their ability to create desired outcomes. When a person believes that certain options are not available to them, this limits what options they think they have for what kind of life

[2] The term "coercion" in a mental health context sometimes refers to forcibly and paternalistically acting upon a person in a way that denies them voluntariness of what happens to them, such as the force and paternalism involved with coercive treatment, involuntary hospitalization, or forcible injection of medication. Here I use the term "coercive" to refer to the way external factors shape desires in ways that are not voluntarily chosen.

they can live by excluding many of the possibilities that they would otherwise like to have. In limiting what options a person believes are available to them, such negative stereotypes reduce the scope of what a person feels like they are capable of desiring and thus reduce the scope of what desires a person can have.

Even when a person does not internalize social norms, the person can still be subject to these norms when the cost of resisting them is too high compared to the cost of following them. A woman living in a culture that has a norm against women working outside the home may desire to resist the norm but find that the cost of resisting – such as humiliation or beating – is too high. To protect herself from negative consequences such as these, she may settle for living according to the norm despite her desire not to. Similarly, a person with mental illness may not endorse the stereotype that people with mental illness are incompetent and needy, feeling that they are indeed competent at least in some areas. However, the cost of appearing competent may be to lose eligibility for disability payments; when there is no guarantee a person can hold down a job over time, this may be too high of a risk to take, and the person may settle for being viewed and treated as too incompetent to work. As Serene Khader notes, sometimes oppressed people have to choose between experiencing well-being through self-subordination (going along with social norms to be accepted by society) and enacting resistance through self-sacrifice (being penalized by society for going against social norms). Sometimes the cost of self-sacrifice is just too high to accept (Khader 2020, 509–510). In this way, norms can limit options for action even when people do not endorse them simply because people feel like they have little choice but to follow them.

How Internalizing Negative Stereotypes Impacts Self-Concept and Self-Regarding Traits

Internalizing negative stereotypes can hinder autonomy not only by unduly restricting options, and thus violating freedom conditions of self-determination, but also by undermining normative authority, which is necessary for the self-authorization aspect of autonomy. The internalization of negative stereotypes undermines normative authority by negatively shaping self-concept, and by hindering the development and exercise of positive self-regarding traits such as self-trust, self-confidence, self-efficacy, and self-respect.

As discussed in Chapter 3, normative authority involves recognizing oneself as the author of one's choices and actions and taking responsibility for what one does. Andrea Westlund describes it as being answerable for one's choices and actions, able to give reasons for them if pressed by others (Westlund 2018, 377). It involves being capable of answering potential criticism about one's choices or actions and being able to justify the action-guiding commitments one holds that underlie one's choices and actions. This reason-giving ability enables a person to see themselves as an authority over what they do and what they value. Reasons that a person could give in justifying their choices and actions must be intelligible to others. As an author of one's choices and actions, a person is able to speak for themselves in regard to these and is answerable for and thus responsible for these in relation to both themselves and to others.

Recognizing oneself as the author of one's choices and actions tends to require having a certain self-concept from which self-regarding traits such as self-worth, self-esteem, self-trust, self-efficacy, self-confidence, and self-respect stem. Having a self-concept of oneself is conducive to having autonomy (Oshana 2005), but not just any self-concept will do. It is important to have a self-concept that has positive effects on self-worth, self-respect, self-trust, self-efficacy, and self-confidence, because these are helpful for regarding oneself as having normative authority, which is seen by some relational theorists as essential for autonomy. When a person internalizes negative stereotypes, however, they develop a negative self-concept that diminishes their self-worth, self-respect, self-trust, self-efficacy, and self-confidence. Their negative view of themselves and their negative self-regarding traits hinder their ability to recognize themselves as the author of their choices and actions and thus diminish normative authority.

Having a sense of self-worth or self-esteem is an evaluative stance that is beneficial for seeing oneself as having normative authority. If a person does not believe they are worthy of receiving moral goods or of interacting with others morally or socially, or if they do not believe their goals and activities have worth, they may not see themselves as having the authority to make decisions for themselves, plan, or act in ways that further their interests. The possession of self-respect is important for valuing one's interests, needs, and goals as worth pursuing and consequently for taking the action necessary to further these. Having a sense of self-trust and self-confidence is significant for developing self-understanding, engaging in self-interpretation, and responding flexibly and effectively to demands put on oneself. Moreover, having self-efficacy, or the belief that one can carry out actions, is valuable to be able to take action in pursuit of one's goals. Self-worth, self-respect, self-trust, self-confidence, and self-efficacy are all highly conducive for exercising autonomy (Benson 1994; Mackenzie 2008, 525).

When a person internalizes negative stereotypes, however, this negatively impacts these self-regarding traits. When they endorse social norms that are considered ideals which they are not able to fully realize, this negatively shapes their self-concept as they feel bad about themselves, and it thwarts their ability to develop positive self-regarding traits. When they cannot live up to ideals that they feel they ought to aspire to, people often feel embarrassed and ashamed, losing confidence, self-respect, and self-worth. When they adopt oppressive norms that, when realized, harm their flourishing and well-being, this can make them feel bad about themselves, too, feeling that they deserve or have no choice but to be resigned to the bads that they aspire to. This, too, damages self-regarding traits and self-concept.

Joel Anderson and Axel Honneth argue that these self-regarding traits are developed through social relationships in which a person learns to see themselves through the eyes of another and through which a person is able to practice the social interactions necessary for developing their moral agency (Anderson and Honneth 2005). They argue that people are necessarily *dependent* on others and thus vulnerable to others for the development of self-esteem, self-trust, and self-respect. Similarly, Benson argues that self-worth is a precondition of autonomy, and self-worth is typically developed through social relationships that teach a person how to regard themselves with respect and esteem (Benson 2005, 110–112; Benson 1994). Damaged self-worth can diminish a person's sense of authority over themselves and thus diminish autonomy. People with mental illness commonly struggle to have positive, healthy social relationships; when they fail to have these, they lose out on the ability to develop a positive self-concept and positive self-regarding traits. When people with mental illness internalize negative stereotypes, and when people around them perpetuate negative stereotypes, this can damage their social relationships and damage their ability to develop a positive self-concept and positive self-regarding traits.

Internalizing the social norms embedded within negative stereotypes can thus impair autonomy in at least two ways. First, it does so by restricting options and thus violating the freedom condition of self-determination. Second, it does so by inhibiting the development of a positive self-concept and positive self-regarding traits, thus undermining normative authority, which is important for the self-authorization aspect of autonomy. The adoption of social norms embedded within negative stereotypes that occurs with internalized stigma harms people in a variety of ways by inhibiting their flourishing, impairing their agency and autonomy, and making them live in ways they wouldn't otherwise want to live.

How the Desires Formed in Response to a Restriction of Options Can Be a Positive Adaptation or a Negative Adaptive Preference

When a person internalizes stigma, this stigma often structures their desires and values in harmful ways. Changing one's desires and values to accommodate restrictions of options can be either empowering or oppressive. When a person develops preferences that align with restrictions in their options, this is sometimes identified as adaptive preferences. When described neutrally, adaptive preferences are preferences a person forms that allow them to adapt to their circumstances. When a person adapts preferences as a way to make the best of a difficult situation, by making choices over what little they can control, developing adaptive preferences can be a positive achievement that empowers the person to have control over a situation that may seem otherwise daunting. But when a person adapts preferences in a way that perpetuates further restrictions of their options or otherwise harms them, adaptive preferences can be oppressive. This oppressive nature is what philosophers usually mean when they refer to adaptive preferences in a pejorative sense.

Positive Adaptation

Adaptive preferences are the preferences a person forms in relation to limited options, where a person chooses something that they would not choose if they had more and better options. The "adaptive" part of adaptive preferences concerns the way desires and choices are formed to adapt to one's circumstances, so that they are formed in relation to the options that are available. The "preference" part of adaptive preferences involves the desires, aims, and choices a person makes in response to those limited options. Preferences are sometimes seen as synonymous with desires, or what a person wants, but they can also be seen as synonymous with values, what a person wants in terms of what they prize and prioritize; aims, or a person's goals and ambitions; or choices, what a person decides to do in light of what they desire and value. Desires, values, aims, and choices are not all synonymous with each other, but the idea of "preferences" in the term "adaptive preferences" can refer to any or all of these. I use these terms interchangeably as synonyms for "preference" depending on the context.

Strictly speaking, adaptive preferences are not always problematic. We adapt our preferences to our circumstances all the time. We always reason under conditions of scarcity and limitation, and we always have to adjust our aims, expectations, desires, and values in relation to what options are available to us. What we choose to do depends in part on what options are available to us. We cannot act without being aware of the limitations brought about by our circumstances and adjusting our mental states and actions to accommodate these limitations.

Not only do we always form desires and make choices in relation to the limitations imposed by our circumstances, but also in many contexts it is admirable and good for us to do so. Adapting to our circumstances is an important component of resilience, or the ability to cope with adversity. Sometimes when we face adverse circumstances, we do not have the power to change the situation to alleviate the difficulty. However, what we do have power over is how we deal with the obstacle. If a person changes their desire for things to happen in a certain way when that way is not a realistic possibility, this change in desire can alleviate the vexation and despair that often accompanies frustrated desires. This kind of acceptance of the reality of what is possible and letting go of desiring what is not possible can lead a person to have a more calm and peaceful disposition, enabling them to deal with their difficulties better (Gosselin 2024, 71–88). Acceptance of the reality of what is possible and letting go of desiring what is not possible is an important way of controlling one's mental states so they fit with reality, allowing them to be helpful and action-guiding rather than leading to frustration and despair.

Adapting our desires and ourselves to correspond with our circumstances often involves learning new skills, capacities, and virtues that reshape who we are, what we value, and what we choose to do. When these changes in what we desire, choose, value, and aim for allow us to grow as human beings, these changes constitute what I will henceforth refer to as "positive adaptation." This is a term used in the psychology literature on resilience to refer to the positive changes a person makes in their values, desires, choices, and behavior in reaction to their circumstances that help them grow and develop as a human being (Gosselin 2024, 22–24; Reich, Zautra, and Hall 2010; Ong, Bergeman, and Chow 2010; Zautra, Hall, and Murray 2010). Being able to adapt in positive ways enables us to flourish even in situations of hardship. When it promotes our well-being and flourishing, adaptation to circumstances – including adaptation of our desires, values, aims, and choices – is good for us. Positive adaptation can be intrinsically good for us insofar as it promotes our well-being and flourishing.

Adapting one's desires and choices to the limitations brought on by one's circumstances is thus an important part of coping with adversity. Not only does adapting one's desires and choices help a person deal with difficult situations but it also provides a source of power for people in situations where they otherwise have little power. When a person has very limited and not very desirable options, and they lack power to change the range of options that are available to them, forming a desire in relation to those options and making a choice among those options can be an important act of empowerment.

Adaptive Preferences

Because adaptive preferences can be a valuable component of coping and resilience, and a source of empowerment in situations where one has little power, it can be difficult to see why adaptive preferences can also be problematic.[3] Yet the term "adaptive preferences" is typically used in a pejorative way to identify problematic desires a person has or problematic choices a person makes.[4] Often these are seen as problematic in that the desires or choices are bad for the person in that they are irrational, nonautonomous, oppressive, and/or diminish a person's flourishing and well-being. As already noted, positive adaptation can allow a person to grow, fostering their flourishing and well-being. Preferences that are adaptive in the pejorative sense, on the other hand, often hinder a person's flourishing and well-being, sometimes by oppressing them, and sometimes by being irrational and nonautonomous.

For preferences to be *adaptive*, they must be formed under conditions when there are limited options, where a person desires something that falls within the available range of what is possible rather than desiring something that falls outside of the range that might be more desirable or better for them. Adaptive preferences are desires for something that a person would not normally desire if they had a wider range of options to choose from, and if this range included options that were better for them. In other words, adaptive preferences are desires that a person has given the range of what is available to the person, where the person would have an alternative desire if they had a different range of possibilities available to them.

[3] For this reason, Dale Dorsey claims that the fact that preferences are adapted is not what is problematic about bad preferences, and thus adaptive preferences are a "red herring" (Dorsey 2017).

[4] Polly Mitchell describes preferences that are adapted to one's circumstances in ways that are helpful to a person's "adapted preferences," distinguishing these from the "adaptive preferences" that are harmful to a person. Rather than using her terminology, which is a bit confusing, I refer to helpful adaptations as "positive adaptation," borrowing from the psychology literature on resilience, in order to distinguish these from the harmful adaptations that "adaptive preferences" consist of (Mitchell 2018).

A person with mental illness might have limited options stemming from a belief that they are incompetent based on the negative stereotype that associates people with mental illness with incompetence. This belief may change their desires about what kind of life they want to live by restricting what they care about and choose to do. Such a change in desire could be seen as either positive or negative. If their perception of their limited options accurately reflects the reality of their situation, and they can find ways to adapt their choices, desires, and values in ways that enhance their flourishing, then these adaptations may be positive. For example, the shift in values that occurs when a person changes their desire for a life partner or children to a desire for a pet, in response to constraining circumstances that make deep, long-term relationships with people seem impossible, can constitute positive adaptation. In this case, the desire for and value of having a pet increases the flourishing and well-being of the person, making their adaptive preference a positive adaptation to their circumstances.

If their perception does *not* reflect the reality of their situation, and/or their ways of adapting diminish their flourishing, on the other hand, then these adaptations constitute adaptive preferences in the pejorative sense. Changing one's desire for and value of having a life partner or children to a desire for and value of living a solitary life can be detrimental to a person when that solitary life is marked by isolation and loneliness. In such a situation, the person's change in desires and values impedes their flourishing and well-being and actually harms them. Thus, such a change constitutes an adaptive preference in the negative sense.

Some of the harmful ways of adapting to limited circumstances caused by an internalization of a negative stereotype about mental illness, such as being incompetent, occur when a person does any of the following:

- Experiencing mental illness symptoms and creating preferences to keep their mental illness untreated, because they feel they deserve to experience symptoms that cause them suffering, because they interpret their symptoms as normal for them so they don't need treatment, or because they don't recognize that they are experiencing symptoms.
- Suffering and forming preferences to live under conditions that make them suffer, because they believe they deserve to suffer.
- Developing a preference to not have children or even pets because they believe they are unable to care for them adequately.
- Forming preferences to rely on others rather than trying to do things themselves because they believe that they are necessarily and inherently dependent and needy on account of having mental illness.

- Creating preferences that reflect lowered ambitions or lowered expectations because they believe they are unable to do or accomplish much.
- Developing a preference to not work in what is perceived as a demanding job because they feel they are unable to carry out the responsibilities associated with the job.
- Forming a preference to avoid socializing with people because they anticipate that they will be too awkward in their social interactions and want to avoid that awkwardness, or because they anticipate that others will judge them, and they want to avoid that judgment.
- Being unable to obtain a job and developing preferences to not work for wages based on a belief that people who do work for wages work too hard and have too much stress in their lives.
- Finding themselves homeless and developing preferences to be homeless based on a belief that they are more free living on the streets; or that people with homes work too hard and have too much stress in their lives in affording and maintaining a home.
- Losing hope and aborting preferences as a result of believing their situation cannot change or their illness will worsen over time.

All of these preferences may be formed or at least reinforced as a result of internalizing negative stereotypes. Some of these arise from a person believing themselves to be incompetent because they have mental illness, and they associate mental illness with incompetence. Other preferences arise from other stereotypes, such as the view that mental illness is permanent and unchanging or that it worsens over time, or the view that people with mental illness are socially inept or necessarily dependent and needy. Internalizing these stereotypes often drives the development of preferences that restrict one's options and limit what kind of life one can live and who one can be. Preferences may be developed in this way whether the internalization occurs through self-stigma, endorsing the stereotype that is applied to oneself, or through attendant stigma, seeing how one can be stereotyped by others and, by paying attention to the stereotype, using the stereotype to inform what one desires, values, and chooses.

In the following sections, I examine some of the ways adaptive preferences can be harmful and problematic for the people who have them by looking at the rationality deficits that they may involve. My discussion refers to the examples already listed of adaptive preferences that people with mental illness may develop when internalizing negative stereotypes about mental illness.

Adaptive Preferences as Rationality Deficits

Many adaptive preferences can be understood as rationality deficits. What makes adaptive preferences wrongful, in these cases, is that they involve problems with reasoning, including creating false beliefs, unduly restricting what constitutes a valid reason, violating the result of a cost–benefit analysis, causing harm that is falsely perceived as benefit, and lack of reflection. Let us examine some of the ways adaptive preferences can be formed through irrational processes.

Self-Deception through the Development of False Beliefs

Some adaptive preferences are problematic because they result from irrational thinking. There are several ways preference formation can be irrational. One way is through the development of false beliefs in the context of self-deception. The development of false beliefs can occur through a process that *appears* rational in that it tries to make first-order and second-order desires consistent with each other. Jon Elster describes this as the "sour grapes" phenomenon, based on the fable of the Fox and the Grapes by Aesop and La Fontaine (Elster 1983). In this fable, the fox originally desired grapes, but upon finding that he cannot obtain them, he changes his view of the grapes so he no longer desires them, in this case by telling himself that the grapes are sour and undesirable anyway. In the sour grapes phenomenon, the change in desire is based on a second-order desire to have one's first-order desires satisfied. When a person is unable to get their desire satisfied, this is viewed as an adequate reason to change one's desire to be something that the person can get satisfied so that their first-order and second-order desires are consistent with each other. However, insofar as this change in desires occurs through the development of a false belief – here, that the grapes are sour, when the fox has no actual reason to think this – it is actually irrational.

In this example, the fox has to engage with self-deception in order to convince himself that a desire he originally had is no longer desirable. He has no *reason* to think the grapes are sour; he merely tells himself that they are so he no longer wants them. Thus he engages in self-deception for the purpose of having the tranquility and peace of mind involved with being able to satisfy his desires rather than having them frustrated. When a person tells themselves that something they originally desired no longer is desirable or never was desirable, they engage in a similar form of self-deception for the sake of changing their desires to fit their circumstances so they can want something they are able to have and thus can achieve a tranquil state of mind. They develop false beliefs, based on bad or absent reasoning, for the sake of making their first-order and second-order desires consistent.

A person who finds themselves homeless and tells themselves that they are freer living on the streets, or that people who have homes work too hard and have too much stress in their lives to afford and maintain a home, probably reasons in this self-deceptive, sour grapes way. They find themselves in a negative circumstance that they would not choose for themselves if they had different and better options, but since they are in these circumstances, they change what they desire to fit their circumstances. If they are homeless, they had better enjoy being homeless, because that is their lot in life. They may believe that there is nothing they can do to change their situation, so it is best to find a way to appreciate, value, and desire it. It is "best" in the sense that if they desire to be homeless, their desires will not be frustrated because they will have what they want. Wanting something that they perceive they can't have – a home – will only bring them frustration and despair, whereas wanting what they do have can give them peace of mind.

When a person tells themselves that they are freer living on the streets, or that people who have homes work too hard or have too much stress in their lives, they are engaging in self-deception because they are telling themselves a story that does not necessarily fit the evidence. They tell this story to themselves because it supports the formation of their desire to prefer homelessness, but it is not necessarily rooted in reality, and it does not reflect what they would want if they had more and better options available to them. They engage in self-deception to make their preference for homelessness seem justified, but when one looks at a wider range of evidence and options, this preference is not actually justified.

Criterial Judgments Based on Feasibility

One way to understand self-deception is in terms of making criterial judgments based on a single factor that appears to outweigh other factors in its importance, rather than making all-things-considered judgments that take all relevant factors into account. Luc Bovens explains the irrationality involved with adaptive preferences in terms of an inconsistency between what he calls criterial judgments and all-things-considered judgments (Bovens 1992). A criterial judgment is the judgment of what a person wants or values based on one specific reason, or one specific aspect of the situation, rather than looking at the situation as a whole. An all-things-considered judgment is the overall judgment a person makes based on *all* (known) aspects of the situation. All-things-considered judgments are relational judgments that are "relative to all of the agent's reasons, i.e., all of the mental states that the agent takes to be relevant to the choice at hand" (Bovens 1992, 70). Criterial judgments can change when an aspect of the situation changes, but all-things-considered judgments generally remain constant because they account for all the possible reasons for action, not just one.[5]

When a person engages in self-deception, the primary factor that appears relevant to them in their preference formation is feasibility of the preference, or the extent to which the preference can be realized (Eftekhari 2021, 73–75). If feasibility is a person's main consideration, then when a preference appears impossible to realize, they will change their preference to be something that instead appears more possible. But this ignores the many other factors that are relevant in making a judgment about preferences. As Seena Eftekhari notes, some of factors that are relevant to determining preferences can include issues of morality and justice, how the action fits within a person's long-term plans or goals, how the action fits within the person's conception of themselves and personal identity, how the action fits the person's conception of the good, to what extent the action will contribute to a person's happiness or well-being, how the action fits with other desires and values the person has, what other people's perceptions would be in relation to the action, how feasible the action is, and to what extent a person would have success in pursuing the action (Eftekhari 2021, 73–75). If a person is being rational, feasibility considerations should not outweigh all other factors when so many different factors are relevant to making a judgment about preferences.

[5] David Zimmerman argues that whether a preference change counts as an adaptive preference or a case of deliberate character planning depends partly on whether the reasons for the preference change apply to all relevantly similar circumstances (a type) or only a specific circumstance (a token). In addition, adaptive preferences involve a downgrade of the preference for an individual token or instance, while deliberate character planning involves a downgrade or upgrade of the preference for a general type. For Zimmerman, then, two considerations are relevant for distinguishing adaptive preferences from deliberate character planning: whether the preference change involves a downgrade or upgrade of the preference (deliberate character planning) or whether it involves a downgrade only (adaptive preference); and whether the preference change applies to all relevantly similar situations (deliberate character planning) or whether it applies only to a specific instance (adaptive preference). Adaptive preferences are less rational than deliberate character planning because they apply only to a specific circumstance and are not generalizable to apply to all relevant situations the way deliberate character planning is (Zimmerman 2003).

When a person who internalizes the stereotype that people with mental illness are incompetent changes their preferences solely in light of how likely it is that they could realize a certain desire or value, they are making criterial judgments based only on reasons of feasibility. This constitutes a defect of rationality because they are not taking other relevant factors into account in their reasoning. Their reasoning is so narrowly focused that it ignores many reasons that should play a role in their decision-making. Preferences are more likely to be harmful to a person's well-being when the reasons given for them are so narrow as to be exclusive of other important considerations, including the good of flourishing and well-being.

Harming Well-Being through Lowered Ambitions

Another way to understand the rationality deficit involved with adaptive preferences is in terms of violating the decisional outcome of a cost–benefit calculation of welfare. Adaptive preferences can lead to reduced overall welfare, or what we can also understand as harms to flourishing and well-being. This might seem counterintuitive. If a person has desires that are unattainable, limited circumstances that prevent these desires from being satisfied can lead to frustration and even despair at the inability to satisfy them. Having unattainable desires that lead to frustration and despair results in diminished well-being. Thus, it would seem that changing one's preferences to be in line with what is attainable would be better for a person, for it would allow a person's desires to be satisfied, leading to an increase in well-being.

However, M. Rickard argues that adaptive preferences actually lead to less welfare because even though the person's desires may be satisfied, what they desire and attain is less than it would be (Rickard 1995, 282). When a person adjusts their preferences in relation to limited circumstances, they typically lower their ambitions, wanting less, or something of lesser quality, than what they originally wanted. It may be possible for their desire to be satisfied, but what they desire is less than what it would have been or what it ought to be. This satisfaction of a lesser ambition can provide less well-being to a person than the alternative, which is to have higher ambitions that are unfulfilled. Having higher ambitions can lead to other goods even when those ambitions are not satisfied; the mere having of these ambitions can be good for a person in various ways. Rickard argues that the good of having higher ambitions, even when unrealized, often outweighs the good of having lower ambitions satisfied.

Adjusting one's preferences in light of constrained circumstances is thus sometimes irrational, if we understand rationality in terms of cost–benefit analysis. When satisfying one's adaptive desires results in greater well-being for a person compared to feeling frustration and despair at not having one's desires fulfilled, then it is rational to adjust one's desires in relation to the limits of one's circumstances, and this shift in preferences one's positive adaptation. However, when satisfying one's adaptive desires leads to a lesser good compared to having higher ambitions that are unrealized, then it is irrational to change one's desires to adapt to constrained circumstances. Many of the preferences listed above that a person may adopt when internalizing the stereotype that people with mental illness are incompetent are irrational in that they lead to reduced well-being and cannot be justified by a cost–benefit analysis.

Harming Well-being through Learned Helplessness and Learned Worthlessness

Another way that adaptive preferences can harm well-being besides lowering ambitions is through adopting a stance of helplessness and worthlessness. Sometimes people change their preferences through learned helplessness and/or learned worthlessness, both of which negatively impact flourishing and well-being (Breakey 2010). Hugh Breakey discusses these as if they are a single motivation for changing one's desires (Breakey 2010), but in fact they can and ought to be distinguished.

In learned helplessness, a person changes their desire because they feel impotent in the face of their circumstances. Learned helplessness was first studied by Martin Seligman (1975). Seligman theorized that when animals that were trapped in a cell continuously tried and failed to get out, they learned that they had no control over what happened to them in their situation, and they developed the belief that their action was futile. He believed that people developed this learned helplessness, too, when they tried and failed to change their negative circumstances. This learned helplessness led people to believe that their action was futile, and they developed a stance of passivity where they did not bother acting because they believed it would do no good (Radden 2000; Rippere 1977; Seligman 1975).[6]

Breakey argues that when people have developed learned helplessness, they may change their desire regarding their situation, since they cannot

[6] One theory of depression is that depression results from learned helplessness, but this only describes some forms of depression; sometimes people engage in self-help behaviors when they are depressed, which goes against the learned helplessness model (Radden 2000; Rippere 1977, 208).

change their actual situation, in order to have more control over what happens to them. Because they feel helpless to try to change their circumstances, they change their desire so that they want the situation they are in. In this case, a person has a second-order volition to have some power and control over their circumstances; their reason for changing their desire is to gain this power. This power is not a power to change the circumstance, but rather a power to change the desire so that they can more easily accept the circumstance over which they have no power.

This constitutes a rationality deficit because it involves a form of self-deception that harms the person's well-being despite their (false) belief that it benefits them. It gives the person a semblance of power but no actual power to change their circumstances, thus making them stuck in a harmful situation. This motivation can make a person accept an oppressive and unjust situation just because acceptance is the only power they have over the situation. Accepting an oppressive and unjust situation can be a way of colluding in one's injustice and can serve to perpetuate existing power relations rather than challenging them. This perpetuates the sense of powerlessness that a person feels over their situation.

In learned worthlessness, a person changes their desire because they feel that their original desire is not worthy enough to act on, or because they feel that they do not have enough worth of their own to have reason to act on what they desire. The lack of worth thus can be found either in the desire itself or in the person doing the desiring. When the desire itself is deemed unworthy, especially by others, a person may decide to desire instead something more worthy, or something that *others* find more worthy. They subjugate their own desire by adopting the desires that others have for them.

When the person sees themselves as unworthy, this undermines their agency entirely, undermining their reasons for acting in general. If a person sees themselves as unworthy, they do not view their interests as something worthy of pursuing, so they do not act on their interests and do not act on reasons that are important to them. They may think that what they desire or value is itself worthy, but they believe that *they* are unworthy to pursue what they desire or value, so the desire goes unfulfilled. In its place, they learn to desire something that they believe better fits who they are as an unworthy person, so they desire something less significant that seems more fitting. In this way, they adapt their desires to a more subjugated view of themselves and a more limited view of what is good for them.

When a person with mental illness develops one of the preferences listed in the section Adaptive Preferences, sometimes this arises from believing that they are impotent over realizing a different preference, one which may empower them or enable them to flourish better. Or this may arise from believing that they are not worthy of pursuing a more empowering preference or a preference more conducive to their flourishing. In either case, learned helplessness or learned worthlessness limits what they can desire and value in oppressive ways. Learned helplessness and learned worthlessness are rationality defects because they involve developing a false belief about what a person is capable of or worthy of – a belief that is much more constraining than it ought to be. Having this false belief harms the person by keeping them in a position of oppression which they sometimes willingly perpetuate themselves. In this way, a person pursues what is harmful to them rather than what is beneficial to them, and thus violates the moral injunction of what a cost–benefit analysis would have them do.

Unconscious and/or Unreflective Preferences

Another way to understand the rationality deficit involved with adaptive preferences is in terms of how reflective a person is about their preferences, either in the course of developing their preferences or at a later time when having the opportunity to reflect on and decide whether or not to endorse the preferences that they have. Ben Colburn argues that desire formation is problematic and thus constitutes adaptive preferences when this process occurs due to "covert influence" (Colburn 2011). Here the desire formation is done as a result of factors outside of the self, and a person is unaware of how those factors are shaping their desires and thus not due to deliberate character planning of their own. Colburn endorses Elster's contrast between negative adaptive preferences, which are shaped unconsciously, and deliberate character planning (Elster 1983, 117–119).[7] In deliberate character planning, a person consciously and deliberately determines their desires and values to be as they wish them to be. They are reflective about the process and make deliberate choices about what to desire and value. This results in positive adaptation.

With adaptive preferences, on the other hand, a person forms their desires in an unconscious, unreflective way. What makes the influence of external factors such as one's given situation "covert," in Colburn's language, is that a person is not aware of *how* the factors influence the shaping,

[7] Elster argues that while adaptive preference formation involves downgrading options that are not available to a person, deliberate character planning involves upgrading options that are available. This process of upgrading can result in the positive adaptation that psychologists praise, in contrast with the adaptive preference formation that philosophers criticize (Elster 1983, 117–119).

even if they are sometimes aware *that* they are shaping; it thus occurs unconsciously. One way that adaptive preferences can be irrational, therefore, is that they are desires that are formed unreflectively and unconsciously.

A person can refrain from reflecting on their preferences either at the time of their formation or at any present moment thereafter. In describing adaptive preferences as unconscious and unreflective, Colburn and Elster focus on the process through which preferences are formed. However, a person could develop preferences in such an unconscious way yet later decide to reflect upon them, and, in doing so, decide whether to endorse the preferences as their own or not. Thus, reflection on preferences can occur at a later time after their formation, and this later reflection and choice of endorsement is adequate for such preferences to be rational. What is irrational is when a person does not reflect on their preferences at any given time, so they never give themselves the opportunity to decide whether to endorse their preferences as their own or not. A rationality deficit occurs when a person fails to reflect on and endorse their preferences independently of how they were formed.

Being unreflective about one's desires and values thwarts the normative authority discussed in the section How Internalizing Negative Stereotypes Impacts Self-Concept and Self-Regarding Traits. It makes a person unable to give reasons for their choices and unable to answer potential criticisms of their choices, thus making the person unable to be answerable for their preferences. Not only does this erode normative authority but it also undermines responsibility, as a person who cannot give reasons for their choices may not be capable of being held responsible for them. Adaptive preferences that arise from being unreflective about one's desires and values thus hinder both autonomy and responsibility, making a person less of a moral agent – and, as a result, making a person seem to fit the negative stereotype of people with mental illness as being less of a moral agent than other people. In a vicious cycle, lack of reflection about one's preferences can make a person seem to embody a negative stereotype, further reinforcing the stigma surrounding them, which they then may internalize, and which can further fuel unreflective and unconscious preferences.

When a person develops any of the preferences listed above as a result of internalizing the stereotype that people with mental illness are incompetent, they often do so unreflectingly. They tend not to be aware that they are forming these preferences in relation to a belief about incompetence; they may be aware that they have these preferences, but not conscious of where the preferences came from. Moreover, they may not choose to reflect upon those preferences at any given time after their formation and so never make the choice whether to endorse them as their own or not. In such a case, a person merely accepts their preferences as they are, without reflecting upon them, and acts on them accordingly. Adaptive preferences that are formed unconsciously *and* not later reflected upon and endorsed constitute rationality defects and are harmful for a person in that they do not necessarily manifest what a person would "really" want if they took the opportunity to reflect upon it.

Conclusion

When people with mental illness internalize the negative stereotypes involved with mental illness stigma, they adopt various social norms that are embedded within these stereotypes, many of which are harmful to themselves. These social norms are oppressive in that they unduly restrict the range of options available to a person, and they contribute to a negative self-concept and diminishment of positive self-regarding traits, including self-worth, self-respect, self-trust, self-efficacy, and self-confidence. These effects of internalizing oppressive social norms impair autonomy and consequently well-being. The internalization of negative stereotypes associated with mental illness stigma often leads people with mental illness to develop adaptive preferences that are harmful to them by diminishing their well-being and flourishing, by oppressing them, and by being irrational and nonautonomous. Internalizing mental illness stigma thus results in grave harms to the person.

While this chapter and Chapter 3 examine the impact of stigma that is internalized through self-stigma and attendant stigma, Chapters 5 and 6 analyze the effects of public stigma that is directed toward people with mental illness by others. The next chapter explains how the self is relational and looks at the way public stigma results in harms of self-constitution (including de-individuation, mis-identification, and dehumanization) and harms of social exclusion. Both of these involve damage to the relational self. The chapter focuses on harms of social exclusion, looking at the way stigma causes moral distancing and outcasting that occurs through separation, status loss, and dehumanization, and ends by examining how these are detrimental to autonomy, agency, and well-being.

CHAPTER 5

Social Exclusion and Harms to the Relational Self

Introduction

Belonging to a community is an important aspect of being human; it is crucial for developing and exercising agency and autonomy as well as for general well-being. Having a sense of belonging is essential for feeling recognized and accepted as a person. It is also important for having one's thoughts, ideas, and feelings acknowledged, validated, and taken up in moral consideration and response, as well as in epistemic discourse and activity. Belonging to a community is critical for being able to make moral and epistemic claims on others, having one's moral and epistemic claims responded to, and responding to others' moral address and epistemic discourse in return. It is also necessary for being able to develop and act on values and goals, and for constructing one's identity. People with mental illness are often excluded from moral, epistemic, and social communities and denied this sense of belonging as a result of being stigmatized against by others through public stigma. Consequently, they are often denied these goods that are connected with belonging.

This chapter begins by examining the nature of the relational self, and how the self is constituted through social relationships and engagement in moral and epistemic practices with others. When public stigma is projected onto people with mental illness by others, people who are subject to that stigma often experience harms of self-constitution. These include de-individuation and mis-identification, as well as separation, status loss, and dehumanization, resulting in social exclusion. Harms related to status loss, dehumanization, and social exclusion severely impact autonomy and agency, and consequently well-being. Social belonging is crucial for the constitution of the self and for autonomy, agency, and well-being; when people are denied this belonging as a result of mental illness stigma that excludes them from the community, they are severely harmed.

Sometimes people are not blatantly socially excluded, but they are uncertain about their belonging status. Stereotype threat – the awareness that negative stereotypes can be applied to oneself – can cause belonging uncertainty, in which people are unsure if they are regarded as members of the moral, epistemic, or social community. Belonging uncertainty brings its own harms, including harms to autonomy. This chapter examines this phenomenon as well.

Let us start by looking at the way the self is relational.

The Relational Self

Feminist philosophers describe the relational self as being *constitutively* relational, where one's self-identity is constituted in part by the person's social relationships (Barclay 2000, 61; Keller 1997; Mackenzie and Stoljar 2000, 7 and 22; Nedelsky 2011; Nedelsky 1989; Oshana 1998; Sherwin 2000). The relational self is socially embedded (Mackenzie and Stoljar 2000, 4), nesting within the social relationships that comprise the person's character and identity as well as the power relations that shape who a person is. In other words, both the social relationships that a person has by virtue of various roles they inhabit – for example, professional, wife, mother, and friend – and the power relations that a person is subject to by virtue of how they are socially situated – for example, gender, class, race, sexual orientation – shape a person's character and identity.

Social relationships impact individuals' character and identity both through the process of moral development and through ongoing adult social interaction. Children develop their personal identity, self-concept, and moral agency through the care and upbringing they receive from caregivers, while adults develop and maintain these through ongoing social interactions they have with specific others. The social interactions a person has with others help to constitute their identity by shaping their experience and how they make sense of their experience through interpretation and meaning frameworks, as well as by affecting their beliefs, desires, and values (Atkins 2006; Meyers 2004).

The formation of identity is also relational, in that how people are regarded by others impacts their own self-concept. When people learn how they are viewed by others, they tend to come to adopt or resist the conceptions of themselves that others have. Often, people internalize others' conceptions of themselves, which influences how they see themselves and consequently what values and goals they have and what reasons for action they find valuable. Even when people resist others' conception of them, others' views still impact their self-concept, their values and goals, and their reasons for action. These in turn all influence their choices and actions, which also help shape who they are.

Moreover, social relationships affect moral development. Both in children and in adults, interpersonal relationships determine what kind of care and concern a person receives, and this impacts moral and character development (Davy 2019). When social relationships provide support and aid, they give people the resources they need in order to develop strengths, skills, virtues, and agency. When social relationships are impoverished, on the other hand, they have a neglectful or even actively harmful effect on personal development, creating a gap between what is needed and what is provided, and leading to fissures in identity and agency formation. Receiving care in a nurturing environment helps a person to flourish as an individual and as a moral agent; not receiving care can have negative consequences on personality, identity, and agency.

In addition, social relationships impact agency and autonomy. People develop values and goals that they then act upon in relation to others. They often learn what values and goals are worth striving for from others, and they tend to develop their own conceptions of these in response to what others find valuable. The development of what reasons for action are worth acting upon is also relational. Sharing one's reasons with others, being answerable for what one does, hearing others' reasons and responding to them, hearing others' perspectives (beliefs, values, and ideas), and making choices about whether and how to incorporate these into one's own perspectives, all occur in relation to others. These processes happen in the context of a moral community where one interacts with others as an equal participant having equal moral status and deserving of equal moral consideration.

What and how a person can know and make meaning of is also formed in relation to social relationships. People generally develop ideas and knowledge through epistemic discourse with others, where they learn about ideas and knowledge that others have acquired, try out ideas on others, and have their own ideas and knowledge legitimized through acceptance of their credibility. They often learn how to make meaning of their experiences and actions in relation to others' meaning systems and in the context of sharing their understandings with others. Being part of an epistemic community where one is seen as having credibility, legitimacy, and epistemic authority is important for a person to be able to participate in the epistemic practices involved with attaining knowledge and making meaning.

When a person has poor relationships or does not really have relationships with others to speak of, or when they are excluded from membership in moral or epistemic communities and their participation is not accepted or recognized by others in the community, they are not able to develop their values, goals, character, identity, knowledge, and ways of making meaning in a robust way. This impedes them from being able to fully exercise agency and autonomy and from being able to develop a rich conception of themselves and a deep moral character. Being excluded from a community leads to an impoverished self-concept, a narrower moral character, and deficient agency and autonomy. This is frequently a problem for people with severe mental illness.

People can experience problems with belonging in a community through several channels. These include various harms to a person's self-constitution, such as de-individuation, mis-identification, and dehumanization. These also include the social exclusion and outcasting that accompanies status loss, dehumanization, and moral distancing, and the belonging uncertainty that accompanies stereotype threat, both of which are explored in this chapter. In addition, problems with belonging occur through the social exclusion and exclusion of participation that arises from discrimination, as discussed in Chapter 6. Through various mechanisms, all of these threaten agency, autonomy, knowledge, and meaning-making, and impoverish self-identity and moral character.

Harms of Self-Constitution

One set of harms that stereotyped people experience are harms to their self-constitution, or harms to the way they are constituted as selves. These include de-individuation and mis-identification, as well as separation, status loss, and dehumanization, resulting in social exclusion. I explain de-individuation and mis-identification briefly, and then I focus more on status loss and dehumanization, which contribute more to moral distancing, outcasting, and social exclusion.

Being recognized by others as equal members of a moral, epistemic, and social community, and as equal participants in the community's moral, epistemic, and social practices, is vital for agency, autonomy, knowledge, meaning-making, identity, moral character, and well-being. When people are recognized as legitimate, authoritative, credible participants in a community's moral, epistemic, and social practices, they have a sense of belonging, which fortifies their sense of self. When people are not recognized as legitimate, authoritative, and credible participants, on the other hand, they are denied a sense of belonging, which negatively impacts their sense of self.

Sometimes a person is not rejected outright by a community, but they are nonetheless uncertain whether they belong. They do not know if they are regarded as a legitimate, authoritative, credible participant in the community's moral, epistemic, and social practices, and they have to act and interact with others in the face of this uncertainty. This belonging uncertainty can be just as damaging to a person and their autonomy as explicit social exclusion. When a person faces stereotype threat, where they see themselves as being able to be judged according to a stereotype by others, or when they experience anticipated stigma and discrimination, and expect to be stigmatized and discriminated against even if these have not actually happened (yet), they experience this belonging uncertainty.

De-individuation and Mis-identification

Stereotypes can create certain harms of self-constitution, namely, de-individuation and mis-identification, regardless of whether the stereotype is negative or positive.[1] The overgeneralization that occurs with stereotyping tends to reduce a person's identity to the stereotyped trait, erasing their individuality and conflating their identity with other generalized identities. This creates harms to an individual's personal identity.

One harm of stereotyping that consists of an affront to a person's identity is de-individuation. This involves not appreciating the individuality of a member of a stereotyped group (Anderson 2010, 45; Blum 2004, 272). One way that de-individuation arises is through the way that stereotypes are reductivist. Stereotypes reduce a person's identity to the trait that is the basis of the stereotype, so that the identity of a person who is a member of the group that is stereotyped is seen essentially in terms of that trait. Other aspects of their identity are ignored and overlooked. Who they are gets reduced to that trait, so their individuality gets erased.

Along with seeing individuals in reductivist ways, stereotyping also dissolves the individuality a person has in relation to others in their social group (Blum 2004, 273). Since everyone who shares that trait has their

[1] A critic may argue that some stereotypes are morally benign and do not always harm individuals. It is very difficult to think of an example where this is the case, because all stereotypes overgeneralize about people and thus reduce people to the stereotyped trait, regardless of whether the stereotype is positive or negative. Perceiving a person in this reductivist way is always harmful to the person, at least as affronts to the relational self as described here.

identity reduced to that trait, everyone who is a member of the group based on that trait is seen as being like everyone else who is a member. The differences between individuals get eroded, so that all members of the group are seen as essentially the same. In reducing identity to the trait that is stereotyped, stereotypes erase individuality and elide differences between individuals who are members of the same group. One aspect of not appreciating the individuality of a member of a stereotyped group thus involves not seeing internal heterogeneity within the group.

Reducing individuals to their stereotyped trait and obscuring internal heterogeneity within a social group are moral harms that occur not only when negative stereotypes are applied to individuals but also when positive stereotypes are applied. Whether the stereotyped trait is seen as complimentary or pejorative, the process of stereotyping reduces the person's identity to that trait, so that only that trait gets noticed and paid attention to, and that trait is all that is seen as mattering in relation to understanding who the person is. The person's identity becomes overgeneralized, and the nuance and complexity of their lived experience and agency become erased.

This overgeneralization and erosion of the nuance and complexity of a person's unique identity often leads people to make assumptions that everyone who has the generalized social trait that is stereotyped is the same. This obscures the internal heterogeneity of the social group. When everyone who has the social trait that is stereotyped is seen in the same way, the assumption is that everyone who has that trait is more similar to each other than different. As a result, differences between individuals within the same social group get eroded. The overgeneralization involved with reducing individuals to a stereotyped trait and the obscuration of internal heterogeneity within a social group both erase the individuality of people who are members of stereotyped groups, regardless of whether the stereotype being applied to them is positive or negative.

Stereotypes of people with mental illness illustrate this. When people with severe mental illness are presumed to be homeless and viewed as dangerous, their apparent homelessness and dangerousness is all that is noticed; other features of their identity are overlooked and dismissed. They are viewed as being the same as all other homeless and dangerous people with mental illness so that differences between individuals are ignored. These moral harms of stereotyping occur with positive stereotypes as well. When people with mental illness are viewed as creative geniuses, their creativity and genius is all that is noticed; other aspects of their identity and experience, including their difficulties with functioning in various life domains, are overlooked and disregarded. Exclusive focusing on a positive aspect of mental illness experience can result in a person's struggles being ignored and downplayed. Moreover, one creative genius is seen as being like all other creative geniuses, so differences between individuals with mental illness are eroded.

These two moral harms that de-individuate a person are related to a third harm of self-constitution, of failing to recognize different kinds of similarities and differences besides those involving the stereotyped trait, so that other ways of grouping individuals are made invisible. Assuming that everyone within a group is the same in relation to a certain generalization makes a person unable to see other kinds of similarities and differences within the group. Without this perception, a person cannot recognize different subgroups of people that fall within the general group (Blum 2004, 275). The harm of not recognizing this is a harm of lack of recognition of people's subgroup membership. Subgroup memberships can be very important to a person, and not recognizing this membership can be a form of disrespect. Blum gives the example of not recognizing the difference between Afro-Caribbeans and African Americans (Blum 2004, 275).

In the context of mental illness, this can involve not recognizing the difference between people who have schizophrenia and people who have dissociative identity disorder (formerly known as multiple personality disorder). People in each of these groups want to be recognized as having the disorder they have, not as having a different disorder that gets confused with it. Not recognizing this difference leads to disrespectful action toward a person who is not seen for who they "really" are, and it can lead to harmful action when others respond to the person in a way that is inappropriate in relation to what mental illness they actually have. This harm of de-individuation is a harm involving lack of recognition of more specific features of a person's identity.

The self is relational in the sense that it is partly constituted by a person's relationships and by their community memberships. When the self is misidentified as belonging to a group it does not actually belong to or as being more alike or dissimilar to others than it actually is, this is an affront to the relational aspect of the self, constituting a disrespect for this dimension of the self. Such disrespect can be fundamentally alienating to a person, making a person feel like a stranger to themselves.

Moreover, the lack of recognition of a person's unique identity that occurs with stereotyping is a moral harm of disrespect. Stephen Darwall identifies recognition respect as one of the important forms of respect that can be given to a person (Darwall 1995; see also Middleton 2006; Thornton

2020). His understanding of recognition respect is based on the identity of a person as a human being, where the person is recognized as having a certain status, namely as having humanity, dignity, and intrinsic value on account of being a human being. Recognition respect is thus the respect a person gives someone when they recognize the other as a human being who is worthy of respect based on having this status (Darwall 1995, 183).

David Middleton identifies a second form of recognition respect that is a recognition of a person's status in relation to others and the community(ies) they are part of (Middleton 2006). When a person is recognized as having status in relation to others, they are regarded as a member of the community, and the social roles they occupy are acknowledged and affirmed. This form of recognition respect is the respect a person gives someone when they recognize the other as a member of the social, epistemic, and moral community, as inhabiting certain social roles, and as an equal participant in the community's practices.

Here I want to suggest a third way that people can be respected or disrespected, based on recognition of their status as a unique individual. When people receive this form of recognition respect, they are seen as having an identity of their own, with their own personal experiences and agency, that make up who they are. When a person's unique identity is not recognized, they are disrespected in a morally important way.

Stereotyping thus harms the relational self in multiple ways. It overgeneralizes about a person, reducing the person to the stereotyped trait; it makes people who share the stereotyped trait seem more similar than they are; it erases differences between individuals who are members of the same social group; and it sometimes misidentifies people as belonging to a social group they are not actually a member of. These affronts to the relational self lead to self-alienation and a lack of recognition respect based on a person's unique identity. These harms to the relational self constitute some of the moral harms of stereotyping.

Moral Distancing, Status Loss, and Dehumanization

Another harm of self-constitution caused by stereotyping and prejudice involves status loss and dehumanization, or seeing stereotyped people as lesser compared to others. As already noted, stereotyping and prejudice lead to a separation and moral distancing of "us" versus "them," where "they" are seen as fundamentally different from and consequently abnormal compared to "us." This separation easily leads to status loss, where those who appear fundamentally different and abnormal are regarded as inferior to those who are "normal." This inferiority often results in viewing people who are stereotyped and prejudiced against as lesser in many ways compared to others, and consequently they are often dehumanized. Separation, status loss, inferiority, and dehumanization all negatively affect how the self is constituted and thus consist of harms of self-constitution.

Dehumanization is the degrading of one's humanity. This can be understood in multiple ways. One way that is very broad comes from Mari Mikkola (2016). Mikkola defines dehumanization as "an indefensible setback to some of our legitimate human interests, where this setback constitutes a moral injury" (Mikkola 2016, 145 and 164). Legitimate human interests are those interests needed to pursue and advance one's goals and aspirations. Mikkola's understanding of human interests comes in part from Joel Feinberg's identification of a "network" of interests that includes physical and mental health, bodily and mental functioning, lack of suffering, emotional regulation, cognitive capacity, lack of anxiety or depression, capacity for social interaction and engagement, financial security, a social and physical environment adequate to get one's basic needs and goals met, and freedom from interference and coercion (Feinberg 1984, 37; see Mikkola 2016, 158). Mikkola contends that a setback to interests such as these that results in moral injury and that is not justifiable constitutes dehumanization.

I think this definition of dehumanization is too broad. A setback to significant human interests that results in moral injury is a good definition for what constitutes a moral wrong, but dehumanization is more specific to the treatment of a person or group of people as lesser compared to other persons, or as having degraded humanity compared to others who are seen as having adequate humanity. Not all moral wrongs constitute dehumanization. For example, when a government dismantles parts of a social safety net – such as by limiting welfare benefits – it surely wrongs the people who need this safety net when it does so indefensibly. It certainly sets back the legitimate human interests a person has in having financial security and can affect other interests by damaging physical health, causing anxiety and suffering, and making one's social and physical environment intolerable. However, such a moral wrong does not seem to constitute dehumanization. It involves treating people who need social safety net badly, but it does not necessarily involve treating them as having degraded humanity or as having less moral status compared to other individuals.

Another way to understand dehumanization involves seeing a person as subhuman and treating them as lesser as a consequence. David Livingstone Smith (2016; 2014) defines dehumanization as treating a person as subhuman, perhaps more like a monster, animal, machine, or trickster. However, he notes that when people are treated as subhuman, it only applies partially. People who are dehumanized occupy a strange metaphysical territory of being seen as human in some ways, such as in the ways they are reacted to, and being seen as inhuman or subhuman in other ways, such as in the kinds of treatment they are seen as deserving. Smith describes dehumanization as a manifestation of the uncanny, which is "creepy, eerie, unsettling, and horrifying" (Smith 2016, 431). The uncanny involves a transgression of metaphysical boundaries, where someone is both like "us" and unlike "us," where "we" are normal, ordinary human beings, and "they" are subhuman in some ways though human in other ways.

For example, people who are viewed as subhuman are still treated as if they are fully human when they are subjected to reactive attitudes, such as when others feel fear, disgust, and blame toward those with mental illness. Peter Strawson notes that we direct reactive attitudes toward people whom we view as *persons*, as fellow human beings and moral agents worthy of being reacted to as having equal moral status (Strawson 2003). When we direct reactive attitudes toward people, we treat them as equal members of our moral community. Nonetheless, people with mental illness are at the same time regarded as subhuman when they are subject to cruel, dismissive, or demeaning treatment that can result from these reactive attitudes.

Individuals can be dehumanized as individuals, or as members of a group which is itself dehumanized. When dehumanization results from stigma and stereotyping, people are dehumanized as members of a group that share the social trait that is stereotyped. The dehumanization of a social group justifies treating its members as being less than human and can be used to rationalize cruel, dismissive, or demeaning treatment. In this way, it seems to justify moral and behavioral discrimination of treating people differently and unequally, in morally worse and unfair ways compared to others who are not part of this group.

Dehumanization as Being Reduced to a Stereotype

Dehumanization occurs as a result of stigma and stereotyping through several mechanisms. One mechanism is through the reductivism of stereotyping. While stereotypes are not always and necessarily false, as explained in Chapter 2, they are always reductive, reducing a person's identity to the social trait of the group they are seen as a member of, and to the stereotyped trait that is generalized about people in this group. With stereotyping, the identity of a person with mental illness, for example, seems to diminish to simply having mental illness, and to whatever generalizations are made about people who have mental illness. This dehumanizes a person by erasing the multidimensionality of a person's identity, reducing them to a trait that is viewed as inherently problematic. In the process, this dehumanizes a person by ignoring the nuance and complexity of their lived experience and identity.

This erasure of the nuance and complexity of their experience and identity is dehumanizing because it degrades important aspects of a person's humanity, including their agency, subjectivity, and identity. When people are reduced to a stereotyped trait, they are seen as less than human, simply the embodiment of a trait. In this way, they are viewed as subhuman. They might be compared to a machine that produces output given certain input or an animal that acts purely on instinct and needs caretaking; these views do not recognize that they have consciousness, experience, and agency. Or they might be compared to a monster that is aggressive, dangerous, and violent, or a trickster that is shifty, untrustworthy, deceitful, and emotionally cold; these conceptions regard the person as lacking rationality, self-control, morality, and responsibility. When a person is considered to be like a machine, animal, monster, or trickster, they are seen as fundamentally different and lesser than other people. They are viewed as being separate from others and experience a status loss through that separation. That status loss is a form of dehumanization.

While others can dehumanize a person who appears to others to be radically different and lesser, a person can dehumanize themselves when they internalize stigma and see themselves as fundamentally different and lesser. When people internalize mental illness stigma, for example, they often dehumanize themselves by reducing their identity to their mental illness. In doing this, they ignore the complexity of their own experience and identity. They come to believe that their identity is entirely wrapped around their mental illness, interpreting all their experience and the actions they consider taking in terms of their mental illness (Tekin 2014; Tekin 2011). In the process, they ignore other aspects of their identity and neglect to treat themselves as multidimensional beings. They may see themselves as being like a machine, animal, monster, or trickster, lacking rationality, self-control, agency, subjectivity, morality, and ultimately an identity that goes beyond their mental illness. This damages their self-concept and self-identity and can easily lead others to view them in the same reductivist way.

Dehumanization as Having Less Intrinsic Value

In addition to reducing a person to a stereotyped trait, thus erasing the nuance and complexity of their experience and identity, a second way that negative stereotypes are dehumanizing is that they often involve thinking of and treating a person as having less moral worth or intrinsic value compared to other people. In his discussion of dehumanization as treating people at least partly as subhuman, Smith starts by focusing on the dehumanization of people perceived to be dangerous or repulsive (Smith 2016, 436). Sometimes they are viewed in this way because of stereotypes, such as Black men who are assumed to be dangerous because of the stereotype that links the social group of *Black men* with the stereotyped trait of *dangerousness*. People with mental illness can also be viewed in this way when they are assumed to be dangerous and violent. People who are perceived as dangerous or repulsive may be regarded as monsters, having less humanity than others and thus being subhuman. This monstrousness and subhumaness gets reinforced when people seem to embody multiple stereotypes that intersect and overlap due to a plurality of social identities that are each dehumanized. For example, Black men who have or seem to have mental illness are seen as doubly monstrous and subhuman, and even more so if they are or appear to be homeless. Such people are viewed as having less intrinsic worth than others on account of their subhumaness.

Smith also broadens his understanding of the uncanny to include people who are viewed as inferior to others, what he calls "enfeebling dehumanization" (Smith 2016, 439). While people who are viewed as dangerous or repulsive are dehumanized as monsters, people who are seen as simple, incompetent, needy, or dependent (as people with mental illness are often viewed) are dehumanized as animals, having less humanity than others in a different way. In enfeebling dehumanization, people are typically compared to animals that are used as means to humans' ends, for example domestic livestock, trained animals, and animals to be hunted. Individuals who are dehumanized in this way are thus treated as having only instrumental value as means to the ends of others who retain their humanity. They are seen as lacking intrinsic value or at least as having less intrinsic value than others.

Like other forms of dehumanization, enfeebling dehumanization can be reinforced through multiple negative stereotypes when individuals are members of multiple groups that are subject to stereotypes that enfeeble humans in this way. For example, Black women who are seen as "welfare queens," indigenous people who are seen as lazy alcoholics, and Latinx people who are seen as menial workers are all subject to negative racial stereotypes that enfeeble them by treating them more like animals to be managed rather than agents of their own right. People with disabilities who are viewed as globally incompetent and needing help with daily tasks are often treated as animals to be managed as well. In these examples, people are seen as lacking the intrinsic value normally afforded to human beings. When such people also have mental illness, they can be multiply enfeebled on account of being subject to multiple negative stereotypes that treat them as animals to be managed and that grant them insufficient intrinsic value and agency of their own.

Other ways that people can be dehumanized as being less than human are by being compared to machines or tricksters. A machine receives input and excretes output and lacks autonomy, morality, and personal identity, all of which are characteristic of being human. When all that matters is the activity, productivity, efficiency, and achievements of a person, they are compared to a machine. When a person with mental illness is regarded as lesser because they are not contributing to society, not being responsible for themselves, and not taking sufficient care of themselves, they are negatively compared to a machine in this way. At the same time, some mental illnesses encourage people to see themselves and others to see the person as a machine. When a person is focused wholly on their output, productivity, and achievement, as sometimes occurs in eating disorders or obsessive-compulsive disorder, they are acting in a way that corresponds with machines. Being seen as being more like a machine than a human diminishes the intrinsic value that a person has, making them seem as if they have less intrinsic moral worth than others.

A person can also be dehumanized by being seen as a trickster, a fairy or other imaginary creature that plays tricks on humans and is thus shifty, deceitful, and untrustworthy. A trickster lacks morality and responsibility, which are also characteristic of being human. Mental illnesses that make a person appear emotionally cold, such as certain personality disorders or autism; and mental illnesses that make a person suspicious, paranoid, and distrustful of others, such as paranoid schizophrenia and other psychotic disorders, can all make a person seem to lack morality and responsibility. Conditions that can make a person act in ways that are untrustworthy and unreliable, such as substance use and impulse control disorders, also make a person seem immoral and irresponsible. In seeming to lack morality, responsibility, and self-control, a person with certain mental illnesses may be compared to a trickster and denied their humanity. In the process, they are seen as having less intrinsic value than others.

Comparison to monsters, animals, machines, or tricksters need not be conscious. Someone can treat a person with mental illness as one of these creatures or things without being conscious that they are doing so, and the treatment is sufficient for it to constitute dehumanization. Intention and awareness is not necessary for perpetuation of dehumanization.

Dehumanization occurs when people are treated as failing to have the capability to possess certain abilities and traits that are characteristic of being human. These include rationality, self-control, agency, autonomy, morality, responsibility, personhood, individuality, subjectivity, and moral status. Sometimes these abilities and traits are diminished as a result of internal factors such as mental illness that create rational, cognitive, or emotional barriers which prevent people from developing, exercising, or expressing these; or as a result of external factors such as the way people are socialized so that they are not given opportunities to develop, exercise, or express these. It is not wrong to view people as having insufficient agency because they have mental illness or because of how they are treated socially if indeed their mental illness or their social treatment creates this loss. In such a case, the moral imperative is to try to increase the agency of people with mental illness to counteract the effects of their illness and/or to correct problematic social treatment.

What *is* damaging to people, however, is when they are seen as being *inherently* defective, as being *incapable* of developing, exercising, or expressing these abilities and traits no matter the intervention. When people are viewed as inherently defective, this is usually because they have a trait such as mental illness that is regarded as an internal flaw. When people are treated as failing to have the inherent capability to possess abilities and traits characteristic of being human – such as rationality, self-control, agency, autonomy, morality, responsibility, personhood, individuality, subjectivity, and moral status – they are treated as being subhuman and as being lesser compared to others: less of an individual, less of a moral agent, less of a person, and ultimately less human. Being lesser in these ways is connected to being seen as having less intrinsic value or moral worth.

The dehumanization involved with seeing some people as subhuman presupposes a moral hierarchy where some people are seen to have less moral worth than others (Smith 2014, 821). Smith says, "Subhuman creatures, then, are creatures that occupy a rank that is lower than the human rank – creatures to which we attribute a lower degree of intrinsic value than we attribute to our own kind and toward which we feel a lesser degree (if any) of moral obligation" (Smith 2014, 821). People who are treated as subhuman are viewed as having less intrinsic value or less moral worth than those who are of higher rank.

Believing that some people have less moral worth than others based on some trait that they share can lead to believing that these people deserve worse treatment because they are seen as being less worthy of moral obligation and respect. This easily rationalizes cruel, dismissive, and demeaning treatment of people who are seen as less than fully human, as lacking intrinsic value and lacking dignity and identity as a human being. People who are seen as having less inherent capability to possess characteristically human capacities and traits such as rationality, self-control, agency, autonomy, morality, responsibility, personhood, individuality, subjectivity, and moral status are easily devalued and treated poorly as a consequence.

People with mental illness are commonly subject to dehumanization, as they are often viewed as subhuman due to lacking characteristically human abilities and attributes (Boysen et al. 2020). Stereotypes of people with mental illness as having less rationality, less self-control, less agency, less capacity, less ability to function, less productivity, less appropriate social interaction, less morality, less capacity for responsibility, and less moral status than others support thinking of and treating people with mental illness as having less worth. When people with mental illness are stigmatized, they are often viewed as inherently defective, incompetent, needy and dependent on others, and/or dangerous and violent. When they are seen as lacking an inherent capability to exercise rationality, self-control, moral agency, and autonomy; or as lacking their own subjectivity and perspective on the world; or as lacking a personal identity (as when they are wholly reduced to their illness); or as lacking morality and responsibility; or as lacking moral status, they are treated as monsters, animals, machines, or tricksters that lack intrinsic worth. They are often treated like creatures or objects that must be managed rather than like moral agents that can determine and pursue the good life for themselves. According to Strawson, when people are regarded as not having sufficient agency, where their actions are not fully their own, they are regarded as objects that can be "managed or handled or cured or trained" (Strawson 2003, 79), similar to domestic livestock, rather than as agents responsible for their own actions.

Self-Dehumanization

While others can dehumanize a person who appears to lack significant human characteristics by regarding them as having less intrinsic value or less moral worth, a person can also dehumanize themselves in this way when they internalize stigma. When people with mental illness internalize mental illness stigma, they often regard their own humanity, or at least aspects of their humanity, as damaged in significant ways by their mental illness. They might see themselves as less rational compared to others, less autonomous, less of an agent, less individuated, less moral, less of a person, or less of a human being. This makes them view themselves as fundamentally different from others, and as lesser compared to other human beings.

When people internalize mental illness stigma, they typically dehumanize themselves by reducing their identity to their mental illness, consequently erasing the nuance and complexity of their experience and identity; and/or by viewing themselves as having less intrinsic value and moral worth compared to others. In both of these ways, a person degrades their humanity, thus dehumanizing themselves. Through the former mechanism, a person downgrades their humanity by eliminating a consideration of themselves as a multidimensional, complex being. Through the latter mechanism, a person views their humanity as already damaged because they see themselves as lacking typical human characteristics such as rationality, self-control, agency, autonomy, personal identity, personhood, morality, and responsibility. This causes them to see themselves as lesser compared to others.

Harms to Autonomy

Dehumanization usually leads to the perception that the people who are dehumanized by being seen as subhuman have insufficient agency and autonomy to be considered full-fledged autonomous agents capable of directing their life, and making and acting on choices based on reasons and values. Their deficient agency and autonomy can be seen as a necessary attribute related to the social trait they are identified with – for example, in assuming that people with mental illness inherently have less agency and autonomy – or as a consequential attribute that results from other aspects of their subhumanness. The quality of being subhuman is often regarded as a global quality rather than a localized one. If a person is seen as lacking *some* characteristics of being human, they may be assumed to lack *many* or *all* characteristics of being human, including having an adequate level of agency and autonomy.

In addition, people who are dehumanized in the sense that the complexity and nuance of their experience and identity is erased also are often regarded as lacking agency and autonomy. If they are seen as not being multidimensional beings, they are not regarded as beings that can make choices for themselves based on reasons and values that they have reflected upon and are invested in. They are not seen as having a multifaceted and complex life that can be self-directed. Being reduced to a social trait and a generalization made about people who possess that social trait erases the power and control a person has over developing reasons for acting that are meaningful to them and in being able to act on their self-reflected choices. A person who is viewed in this reductivist way is seen as having deficient agency and autonomy, and consequently they are often not given opportunities to be able to develop and exercise what autonomy they have.

Being seen as having less autonomous agency is self-reinforcing, causing a person to actually have reduced autonomous agency. When a person is seen as having less autonomy, they are given fewer opportunities to develop and exercise their autonomy, and so their autonomy diminishes. When they are seen as not having power and control over their actions, they are subject to more barriers that reduce their power and control. Autonomy is causally relational in that a person has to be treated as a member of a moral, epistemic, or social community in order to be given opportunities to be able to practice and hone the autonomy skills that constitute autonomous agency. When a person is perceived as not being an autonomous agent, they are denied these opportunities and not treated as a member of the community, preventing them from developing and exercising their autonomous agency.

Moral Distancing, Outcasting, and Social Exclusion

The separation and status loss that stigma and stereotyping create constitutes what moral philosophers call moral distancing. Moral distancing involves seeing people who share a certain trait as being *more* different from oneself than others who do not share that trait so that difference is seen as an unbridgeable gap that makes the stereotyped group of people "other" than oneself and demarcates a "them" versus an "us." Moral distancing as well as dehumanization support outcasting. People who are similar to oneself in relevant ways are seen as belonging to the community, whereas people who are seen as fundamentally different are seen as not belonging; they are outcasted.

Outcasting consists of socially excluding people from the community. As a consequence, they are prevented from being able to participate in the social, moral, and epistemic practices of the community, where members recognize each other as legitimate participants in these practices. The more that people are viewed as different and lesser compared to others, the more outcasted they are.

A person can be multiply outcasted when they are subject to numerous negative stereotypes and seen as lesser in several ways. People who belong to more than one minority social group – such as being Black, being a person of color, having physical disabilities, being homeless, and/or using drugs – are subject to several negative stereotypes at the same time and are afforded less status in multiple, intersecting ways. They are socially rejected on account of many aspects of their identity and thus can be said to be *more* outcasted than those who belong to only one minority social group by virtue of their identity.

Outcasting people who are seen as not belonging to the community can take many forms. Often it involves shunning and avoiding people through both physical avoidance (such as crossing the street so one does not have to walk by someone who appears homeless and mentally ill) and interactional avoidance, where one finds ways to avoid having to interact with people who are seen as fundamentally unlike oneself. This can lead to profound social isolation. Outcasting people by refusing to engage and interact with them is a way of morally and behaviorally discriminating against them, or treating them differently and unequally in a negative, unfair, and worse way. Moral distancing rationalizes moral and behavioral discrimination by seeming to justify unequal, unfair, and worse treatment of people on account of their being radically different.

In addition, outcasting people can also involve passing judgment on those people by shaming and blaming them for what are interpreted as bad actions. As discussed in Chapter 3, people who are subject to negative stereotyping often have negative, suspicious, or malicious intentions ascribed to them so that their actions appear bad and ill-willed. When a person is outcasted, they and their actions are typically viewed through a negative lens so that everything they do appears morally bad and socially problematic. This negative lens further rationalizes moral and behavioral discriminations. Since a stereotyped person appears to act badly, this seems to justify treating them in different and worse ways compared to others.

Mental illness stigma causes significant social exclusion: People with mental illness report frequently experiencing negative attitudes directed toward them and prejudicial actions resulting in social exclusion (Angell, Cooke, and Kovac 2005; Dinos et al. 2004, 119; Mezey et al. 2022; Reinhard et al. 2020; Thornicroft et al. 2022, 1447). Stigma contributes to a variety of forms of social exclusion, including family estrangement and increased divorce (Adu et al. 2023; Farrelly et al. 2014); prejudice and discrimination in employment, housing, and healthcare (as discussed in Chapter 1); as well as shunning and avoidance in public social settings. Discrimination also prevents people with mental illness from participating in normal activities of society. The social exclusion of people with mental illness inhibits people from engaging with others in a moral and epistemic community as equal agents participating in moral and epistemic practices with each other. This erodes people's sense of belonging within a community; it disables them from developing and exercising the reasoning, valuing, and choosing capacities involved with the self-governance aspect of autonomy; it limits what options are available to them, diminishing the opportunity conditions of the self-determination aspect of autonomy; and it deflates their personal identity as they are prevented from envisioning who they want to be and how they can try to become that person.

Self-Imposed Social Exclusion

While actual stigma outcasts people as not belonging to society, anticipated stigma makes a person *feel* outcasted. A person may not have been actually stigmatized by others, yet they may anticipate being stigmatized because they are aware that they share a trait that is stereotyped and aware that stereotypes involving that trait exist. This awareness makes them anticipate that they will be stereotyped and stigmatized by others even before anyone actually does stigmatize them. This anticipated stigma makes a person feel outcasted before anyone actually treats them as an outcast. It makes them feel fundamentally different from others and makes them assume that they will be rejected by others as not belonging to the community.

Oftentimes this leads people to outcast themselves preemptively before they are outcasted by others. Even though self-outcasting leads to the same social isolation that being outcasted leads to, it can feel empowering in a context where a person otherwise has little agency. It gives a person power over their fate instead of being subjected to the power of others. Many people who share a stereotyped trait socially isolate themselves to protect themselves from being rejected by others by socially withdrawing from the community. They discontinue participating in moral, social, and epistemic interactions with others, withdrawing from the community in which such activity is practiced. People who have mental illness often isolate themselves from others, anticipating that they will be rejected by others on account of having mental illness, and wanting to protect themselves from that rejection by withdrawing socially before they can be outcasted.

Moreover, the self-dehumanization that occurs with internalized stigma further motivates people to put moral distance between themselves and others, and to withdraw socially from engagement with others and to isolate themselves. When people with mental illness internalize mental illness stigma, they often see themselves as a creature or object that has less intrinsic worth than other people, and thus deserves to be treated poorly and cast out of the community as a result. They often use this belief to rationalize treating themselves poorly, such as through self-harm, suicidal ideation, not engaging in mental health treatment, and social withdrawal. They may believe that they deserve such poor treatment by virtue of being subhuman.

In addition, they may believe that others will judge them as poorly as they judge themselves, where others will see them as fundamentally different and lesser than other people. This may lead people with mental illness to believe that others will treat them poorly, such as through discrimination, and that they will be outcasted because they are seen as being too different and lesser to belong to the community. When they believe they will be discriminated against, this can cause anticipated discrimination, where they fear they will be discriminated against in a variety of contexts, and they have to react to that potential discrimination in all their choices and actions. This can lead people to act defiantly in trying to resist discrimination, but it can also lead to resignation, fear, and anxiety, causing people to lower their ambitions and expectations so they will not be disappointed through rejection and discrimination.

Internalized stigma and self-dehumanization provide further motivation for a person to outcast themselves preemptively, withdrawing socially from the community so that they no longer engage with others. Having a sense of belonging in a community requires feeling similar enough to others that one is seen as an acceptable and legitimate member of the community. When a person feels fundamentally different from and lesser compared to others, this destroys their sense of belonging as they feel unacceptable, too illegitimate to be a member of the community. Rather than trying to prove their acceptability to others, they often withdraw preemptively before others can deem them unacceptable. This gives them power over an experience that is otherwise very disempowering, but it creates deep social isolation.

Self-dehumanization through internalized stigma thus supports moral distancing and self-imposed social exclusion in many ways when a person treats themselves as being lesser compared to others and expects that others will treat them as lesser, too. It often motivates a person to treat themselves badly, worse than others. It can cause a person to believe that they deserve to be treated badly by others, which can allow them to accept poor treatment of themselves rather than trying to resist it. In addition, self-dehumanization can lead a person to anticipate that they will be treated badly by others, causing them to lower their ambitions and expectations. Finally, self-dehumanization motivates and supports self-outcasting and social withdrawal, and consequently social isolation.

How Social Exclusion Diminishes Autonomy

The moral distancing and social exclusion involved with stigma and stereotyping – including actual stereotyping and stigmatization, anticipated stigma, and internalized stigma – cause many harms to individuals, including the loss of agency and autonomy. Agency and autonomy are relational sets of capabilities and skills that require participating in a moral, epistemic, and social community with others to be able to develop and practice these capabilities. In other words, a person needs to be able to engage in social interactions with others in order to practice the skills and capacities involved with agency and autonomy. These include skills such as reasoning skills; emotional regulation and appropriate emotional response; forming and testing beliefs; considering and weighing evidence; revising beliefs, desires, and values in relation to evidence and reason; choosing; and valuing. In addition, Diana Tietjens Meyers identifies several skills necessary for autonomy, including introspective skills that allow a person to self-reflect; imaginative skills that help a person consider their self-concept; memory skills that aid people in connecting past, present, and future; communication skills that assist people in learning from and sharing with others; analytical and reasoning skills that allow a person to consider and choose different self-concepts; volitional skills that enable a person to pursue the self-concept they choose for themselves; and interpersonal skills that help a person resist oppressive social norms (Meyers 2004, 31). All of these skills and capacities require individuals to be able to participate in the moral, epistemic, and social practices of a community where they are recognized as equal, legitimate, credible, and authoritative members.

As relational autonomy theorists point out, autonomy is at least causally relational: people develop autonomy in the context of social interactions and social relationships by participating in moral and epistemic practices with others within a shared moral and epistemic community (Friedman 2003, 15; Mackenzie and Stoljar 2000, 22; Meyers 2004). When people are not treated as fully participating members of a community because they are excluded from the community or discriminated against, they lose opportunities to engage in moral and epistemic practices and thus to develop and exercise autonomy capacities involved in making choices for themselves and in directing their life as they wish. This can prevent a person from adequately developing the capacities required for the self-governance aspect of autonomy, infringing on the competence that is required for autonomy.

When people are outcasted and/or discriminated against, or when they socially withdraw from engagement with others, they are prevented from having opportunities to develop and exercise the skills and abilities involved with the self-governance aspect of autonomy. This reduction in opportunities from developing and exercising agency and autonomy leads to an erosion of competence and thus an actual loss of self-governance and self-direction as people are not able to practice the skills and abilities associated with these. The diminishment of autonomy also impacts a person's personal identity as they may become unable to envision what their commitments and values are and who they want to be, and unable to plan and work toward personal development and becoming who they want to be, different from who they currently are.

Social exclusion and social isolation also limit what opportunities are available to a person when they are deciding between different options to choose from, thus diminishing the opportunity condition of the self-determination aspect of autonomy. When people are socially excluded from a community's practices, they have fewer opportunities to pursue various goals they may have. They may be denied access to goods such as education, healthcare, employment, or housing. This puts major constraints on what options are available to them, making the range of choices they have much narrower. When multiple options are denied them, this can even force them into being able to make only a single choice, one they would not choose if they had other options available to them. This can severely diminish their ability to be self-determining, thus greatly impairing their autonomy.

Belonging Uncertainty of Stereotype Threat

Stigma creates other mechanisms that threaten a person's sense of belonging and their capacities for self-governance besides dehumanization, moral distancing, and outcasting. One of these mechanisms is stereotype threat, which is discussed in Chapter 3. According to Paul Benson, stereotype threat is relational in at least two senses. First, stereotype threat involves an awareness of one's potential to be judged by others, making a person see themselves through the eyes of others (Benson 2015, 131). A person engages in self-appraisal about their own capacities, competence, and performance in relation to the stereotype and others' perspective of them, and they become self-conscious and aware of their anxiety and stress level. Stereotype threat would not exist if a person were unaware of others' perception of them, or if there were no others to perceive or judge them. It is only through others' perception and potential judgment that stereotype threat occurs.

Second, stereotype threat involves a self-identification of belonging to the group that is being stereotyped (Benson 2015, 131). Only because a person sees themselves as a member of that group does stereotype threat have any power. In the context of mental illness, this means that a person has to self-label as having mental illness, identifying mental illness as a relevant and salient feature of their identity. If a person did not see themselves as a member of that group, there would be no basis for thinking that a judgment regarding that stereotype would apply to them. A person who does not self-label as having mental illness, therefore, would not believe a stereotype regarding mental illness could apply to themselves and so would not experience stereotype threat. Such a person would not be self-conscious about how they were perceived. Stereotype threat exists because of the sense of belonging and identity a person has with the group that is being judged.

Stereotype threat risks exposing a person to a deep sense of shame based on who they perceive themselves to be in relation to how they believe others perceive them. Shame is a negative feeling toward oneself based on who one is. A person who has a mental illness who is not subject to stereotype threat and who does not believe that negative stereotypes about mental illness apply to them may not feel shame because they feel more comfortable with who they are in relation to the mental illness. Without the connection to a negative stereotype, having a mental illness is not a trait that triggers shame. However, when a person is worried that they will be judged negatively as fitting a negative stereotype, or when they themselves believe that a negative stereotype about mental illness applies to them, then they feel like a trait which is unavoidably part of them makes them a lesser human being. This triggers the negative feelings toward themselves that constitute shame.

Benson notes that people who are subject to stereotype threat often want to feel like they belong within dominant groups but often feel uncertain about their sense of belonging (Benson 2015, 134). This "belonging uncertainty" occurs when, as Gregory M. Walton and Geoffrey L. Cohen note, "members of socially stigmatized groups are more uncertain of the quality of their social bonds and thus more sensitive to issues of social belonging" (Walton and Cohen 2007, 82). Belonging uncertainty does not necessarily involve a *belief* that one does not belong; it is more like a *hypothesis* that one does not belong, a hypothesis which awaits evidence for confirmation or disproof (Walton and Cohen 2007, 83). People who have belonging uncertainty interpret their own difficulties as evidence that they do *not* belong because this seems to indicate that they lack a trait that is seen as salient to determining belongingness.

If intellectual skill, coping, competence, independence, and normalcy are important traits for belonging, then struggling with challenges seems like evidence that one lacks these traits, and thus lacks an important aspect of belonging. People who do not have belonging uncertainty – who feel secure in their sense of belonging and in the social bonds they have – do not view their challenges as evidence of not belonging, because they do not question whether they belong. When a person who has belonging uncertainty struggles to prove that they are competent, intelligent, independent, and normal, on the other hand, they see their struggle as evidence that they lack a trait that is important for belonging and thus believe that this confirms their hypothesis that they do not belong. This can make them feel more isolated, judged, and shamed.

The uncertainty involved with belonging uncertainty is an uncertainty about how they will be judged – whether they will be seen as fitting the negative stereotype or not. The uncertainty creates a different kind of anxiety than the anxiety associated with *knowing* that they are being judged as fitting a negative stereotype. Both knowing that they will be judged so and being uncertain about whether they will be judged so can be detrimental to the person. Knowing that others perceive themselves in terms of a negative stereotype can make a person angry, resistant, sad, demoralized, ashamed, and embarrassed, as well as anxious about how others will react to them based on others' judgment of them. Knowing that they are perceived as confirming a stereotype leads to many different kinds of negative reactive attitudes, some of which may be directed toward those who judge, but others of which are directed toward themselves. Having certainty about their lack of belonging can lead a person to significant negative self-judgment and shame about their very existence.

Not being sure whether others perceive themselves as confirming a negative stereotype, on the other hand, gives a person great anxiety because they want to avoid having to feel that self-judgment and shame, and they know that there is a possibility that they *can* avoid it if their behavior is seen as not confirming the stereotype. This anxiety pushes the person to work hard to try to disprove the stereotype, causing stress and leading either to overwork or to disengagement if a person gives up on trying to fight what may seem unsurmountable. All of these responses can undermine the person's competence, motivation, effort, and performance – and consequently undermine the competence and authenticity conditions of self-governance, as discussed in Chapter 3.

When a person has this sense of not belonging to the dominant group, this affects them in one of two ways. It can drive them to seek out more connection and support with people of the marginalized group to which they belong – such as those with mental illness – to make up for the lack of connection and support they receive from the dominant group. When this happens, their self-identification as a member of the stereotyped group becomes even more important to them because it is their only, or one of their only, source of connection and support. This stronger self-identification as a member of the stereotyped group makes them even more vulnerable to stereotype threat, because the trait that is stereotyped becomes even more important to them as a source of identity and belonging. Thus, stereotype threat is self-perpetuating: It can make a person feel as if they do not belong to the dominant group, even as they want to belong, which can drive them to seek out connection and support more with the stereotyped group, making them more vulnerable to stereotype threat.

Alternatively, the sense of not belonging to the dominant group can make a person feel as if they do not belong *anywhere*, making them feel more isolated and alone. If they do not see the stereotyped group of which they are a member as a source of connection and support, then they will not be motivated to connect and self-identify with it. This can lead the person to be less vulnerable to stereotype threat, as not identifying with the group that is stereotyped can alleviate the person from feeling like other people will think the stereotype applies to them. If they do not think the stereotype applies to them, and they believe that others will recognize that the stereotype does not apply to them, then they will be relieved of the burden of feeling like they are being judged by others.

On the other hand, if the person does not themselves feel like the stereotype applies to them because they do not self-identify with the group, but if they still recognize that *other people* will think the stereotype applies to them because *other people* identify them as a member of the group, the lack of self-identifying with the group will not change anything.

It is the fear of being judged by others that drives stereotype threat, not the endorsement of the negative stereotype as true or as applying to oneself. As long as a person is aware that *others* perceive them as identifying with the stereotyped group, they remain vulnerable to the potential of being judged by others.

How Belonging Uncertainty Impedes Autonomy

Benson points out that belonging uncertainty is another avenue through which stigma interferes with autonomy (Benson 2015, 134). When a person is unsure of whether they belong, wanting to belong but feeling as if their sense of belonging is tenuous or threatened, this affects the same psychological processes that stereotype threat affects, thus decreasing motivation, effort, and performance. Consequently, this decreases competence, which enhances feelings of being different or abnormal and makes a person more vulnerable to being dependent on others. The anxiety, stress, and overcompensation involved with belonging uncertainty also reduce cognitive, emotional, and volitional capacities, which damages self-understanding, self-reflection, and self-appraisal. These impairments violate both competence and authenticity conditions of self-governance.

Benson argues that although a person might not have a sense of belonging to a dominant group when they are identified as belonging to a stereotyped marginalized group, they can still retain *some* sense of belonging if they are able to maintain the ability to view themselves as being able to be answerable for their actions by giving reasons for their choices and addressing potential challenges. He describes this as "having a place as a reason-giving agent among other reason-giving agents" (Benson 2015, 136) and argues that this is a necessary condition for autonomy, namely the self-authorization aspect of autonomy. But whether a person sees themselves as capable of being answerable for their action depends in part on how others perceive them and whether they feel a sense of belonging to a relevant group of others with whom they can be answerable. He says that a person's attitude toward their social belonging affects (though it does not completely determine) how they view their sense of "worthiness to answer for their decisions and actions" (Benson 2015, 136).

Their attitude toward their own social belonging is formed in part by the reactions they experience from others in trying to belong. When others accept them as a member of their community, it is easier for them to feel the sense of social belonging that is required for seeing themselves as answerable to others. When others reject them, however, it is harder for them to feel this sense of social belonging and thus harder to see themselves as answerable for their actions. A person who is seen to be part of a group that is negatively stereotyped may experience significant rejection from members of dominant groups that are in a position to judge them. The potential for rejection can be strong enough to make a person feel already rejected before they are even judged. Because rejection from the dominant group is possible or even likely in cases where stereotype threat applies, it is incumbent on members of stereotyped groups to seek out a sense of belonging with *other* people who are not part of the dominant group.

In many cases, this involves seeking out a sense of belonging with others who are part of their same stereotyped group. Because they share the experience of potential rejection based on their identity with this group, they have an automatic basic understanding of each other's situation, which can be the springboard for empathy, compassion, connection, and support. To counter the potential rejection from members of the dominant group, therefore, self-identification with the stereotyped group can be an important source of acceptance and all the goods that acceptance brings, including a sense of social belonging.

In some cases, potential rejection can lead a person to seek out a sense of belonging with others who are marginalized in different ways, who are in different marginalized groups other than their own. This can be an important source of bridging connections between different groups of people who are stereotyped and marginalized in different ways, but who nonetheless share the status of being stereotyped and marginalized by dominant society. Both of these sources of connection – connecting with others who are within the same stereotyped group that is the basis for the stereotype threat one is subject to, and connecting with others who are in different stereotyped and marginalized groups who may face their own kinds of stereotype threat – can be crucial for a person to feel the sense of belonging that they need in order to be able to see themselves as answerable for their actions to others.

Conclusion

Having a sense of belonging and being accepted as an equal member of a community is crucial for individuals' self-constitution, identity, agency, autonomy, and well-being. Being regarded as a legitimate, credible, authoritative member of one's social group enables a person to develop and exercise the skills and capacities necessary for autonomy, and being seen as a member of the community gives people the support of others who can help the person navigate difficulties, including difficulties caused by stigma. Being treated as a multidimensional, unique individual helps a person develop a self-concept that respects their complexity, and being viewed as a full human being respects their dignity and worth. In endangering all of these ways of being recognized by others, stigma threatens the constitution of the relational self. Consequently, it is paramount for people who are stigmatized on account of mental illness to be accepted and welcomed into the moral, epistemic, and social community as an equal member and a full human being with all the complexity and nuance that individuals possess.

This chapter deals with harms to the relational self, including the constitution of the self, social exclusion, and belonging uncertainty. These harms result from both public stigma that is projected onto people with mental illness by others and internalized stigma that people with mental illness impose upon themselves. The next chapter addresses harms that are caused specifically by public stigma aimed at people with mental illness by others, namely the harms involved with discrimination.

CHAPTER 6

Harms of Discrimination and Microaggressions

Introduction

Public stigma that is directed at people with mental illness by others often takes the form of discrimination. Discrimination is a broad concept used in several different contexts. Epistemic discrimination involves making distinctions among an array of objects, while moral discrimination consists of making moral distinctions between various people. Behavioral discrimination concerns acts where some people are treated differently from and worse than others. As an epistemic practice, epistemic discrimination is a morally neutral concept that does not necessarily involve or lead to treating some people differently from others. When we use the term "discrimination" in a moral sense to indicate differential treatment, on the other hand, we are referring to the moral discrimination of thinking that some people are distinct from others and the behavioral discrimination of treating these people in a disparate way compared to others in our actions.

When I use the term "discrimination" in this chapter, I am referring to behavioral discrimination that stems from the moral discrimination of thinking of some people as different from others, thus seeming to justify differential treatment. Specifically, this involves discrimination that is based on the possession of socially salient social traits such as having mental illness. Moral discrimination, or thinking of some people as distinct from others based on a social trait they possess, is the type of judgment that is involved with stereotyping, as explained in Chapter 2. Behavioral discrimination involves treating some people differently from, unequally compared to, and worse than others based on them possessing this social trait.

Behavioral discrimination typically begins with attributing negative stereotypes to people who possess the social trait so that they seem to be inferior in some way, based on the idea that their social trait violates important social norms that are seen as worthy of being upheld. When people have a social trait that seems to violate these norms, this can seem to justify treating them in worse ways than others, because their norm violation makes it so that they don't seem to deserve to be treated the same as or as well as others. Wrongful discrimination is behavioral discrimination that is unjustifiable, and what makes it unjustifiable, as I argue in this chapter, is the various harms it causes. Wrongful discrimination thus involves treating people who possess the socially salient social trait *worse* than others by causing harms to them that others don't experience.

A more subtle form of behavioral discrimination consists of microaggressions, which are speech and action that insidiously devalue, discredit, disrespect, and demean a person. When people are the target of microaggressions, they experience some of the harms that result from discrimination, including being demeaned and being subject to agential harms of experiencing a loss of deliberative freedom and diminished autonomy. While this chapter focuses on the harms of discrimination, the harms of microaggressions are discussed as well.

As indicated in Chapter 1, people with mental illness are commonly discriminated against in many areas of their lives, including employment, housing, healthcare, social activities, and just generally being in public. They are treated differently from and worse than others. This chapter examines the harms involved with discrimination that people with mental illness experience, showing that discrimination impacts their circumstances, their resources and opportunities, their options, their agency, their autonomy, and their well-being. I argue that the wrongfulness of discrimination consists in its many harms. My approach is pluralistic in that there are multiple aspects of discrimination that can make it wrongful, because there are multiple ways that it can be harmful.

Discrimination and Power

Discrimination is an act of power. Power is the ability to affect outcomes, so discrimination is an act that affects the situation of people who are discriminated against. One way of affecting outcomes involves making a person's situation better or worse. Wrongful moral and behavioral discrimination makes the situation of people who are discriminated against worse by causing them various kinds of harms.

While epistemic discrimination involves the epistemic power of exercising epistemic agency through making distinctions, moral discrimination is an act of moral power where a person exercises moral agency through making and ultimately acting on moral judgments. Moral discrimination consists of thinking differently of some people compared to others. This involves making moral judgments about the moral quality of the person one is judging or of their actions, as well as making a normative comparison of the person or their actions to other people and the actions of others. Making moral judgments tends to consist of assessing a person or their actions as good or bad, and making moral comparisons to the goodness or badness of others and others' actions. Moral discrimination requires having the moral authority and epistemic credibility to make judgments that are taken seriously and responded to through uptake and responsiveness. Having the moral power to make normative evaluations in turn requires having the authority and credibility involved with having power.

Behavioral discrimination consists of acts of treating some people differently from others based on a social trait they share, sometimes backed up by policies that regulate this differential treatment. Behavioral discrimination involves exercises of agency and power. Some of the powers that people exercise when they behaviorally discriminate include the power to make choices in how to treat people, the power to act, and the power to exert control over what happens to a person. In order to have this power, an agent has to have the authority to act according to their choice, the authority to be able to choose to do a particular action over other actions they could choose instead, the authority to shape outcomes that affect others, and the credibility for their action to be taken seriously and responded to. (Taking an act seriously does not mean approving of it. When people respond to discrimination with indignation, disapproval, anger, defiance, or frustration, they are taking the act of discrimination seriously enough to bother responding to it.)

Power is relational in that some agents have more power, while other people have less. When people have more power, they can often direct this power toward determining the outcomes of those with less power. Thus, those who are socially situated in ways that grant them less power are subject to and easily subjugated by the power of those who have more. Discrimination occurs in the context of power relations in which some people are situated in ways that give them the power to make decisions that affect others and to make choices about treating different groups of people differently. Those who are discriminated against have less power and are situated in ways that allow their situation to be shaped by the actions of others based on how others treat them. When discrimination occurs, discriminators treat some people in different, unequal, and worse ways compared to others, impacting their situation in detrimental ways.

As an act of power, discrimination has the potential to shape what happens to the person who is discriminated against. How a person is treated affects what resources and opportunities the person has access to, and what obstacles they have to overcome to be able to act as they desire. By impacting what resources and opportunities a person has access to, and what options and limitations a person faces, how a person is treated affects the ways in which the person's agency is enabled or disabled. Such treatment also affects what kinds of choices a person can make, what they can value, what reasons for action make sense to them, and what action they can take in relation to their reasons, values, and choices, thereby shaping what kind of autonomy they have. When a person is treated with discrimination, this creates harms that obstruct their choices and actions and prevents access to resources and opportunities that would otherwise be available to them. This constrains their agency and autonomy, and consequently their identity, by limiting what they can do and who they can be. Discrimination thus creates great harms for people who are discriminated against.

Some of these harms constitute the distributive and structural injustices discussed in the Introduction and Chapter 1. In diminishing the resources and opportunities that a person has access to, discrimination creates distributive injustices of preventing people from having access to the goods they require to get their basic needs met and to achieve standards of justice. Such standards of justice include (depending on the theory of distributive justice referenced) preference satisfaction (Arneson 1989; Dworkin 1981a); living a life of equal desirability (Arneson 1989; Dworkin 1981a); living a life based on one's own conception of the good (Dworkin 1981b); having the opportunity to hold social positions and offices (Rawls 1971, 90–95) and to fulfill social roles (Reader 2007, 55); having agency and autonomy (Brock 2005; Copp 1992); having one's human rights fulfilled (Kelly 2004); and being able to live a worthwhile or even minimally decent life (Pogge 2008, 43–44).

In setting limitations on what options for action a person has, discrimination also creates constraints on a person's action so that their agency and autonomy are reduced. When discrimination is based on how a person is socially situated within social structures that benefit some who are treated as more morally worthy and burden others who are treated as having less moral worth, this constitutes structural injustice. Structural injustice subjugates those who are discriminated against to those who do the discriminating, creating an unjust balance of power that grants some people power and authority over others.

Discrimination harms people with mental illness in numerous ways. Many cases of discrimination create material disadvantage that affect a person's functioning, well-being, and quality of life. Sometimes discrimination consists of distributive injustices that harm by preventing a person from having access to the same resources and opportunities that others have access to (or that they would have access to if they were not discriminated against), for no morally salient reason; or to basic goods required for equal participation in society. Discrimination can also unfairly subordinate a person through structural accommodations that discount their needs and interests and censure them. Furthermore, both discrimination and microaggressions demean a person by degrading them and treating them as having less moral worth compared to others. In addition, both discrimination and microaggressions lead to agential harms, including a loss of deliberative freedom, and the harm of diminishing autonomy. Let us examine these different kinds of harms caused by discrimination and microaggressions.

Harm as Material Disadvantage

One type of harm that people with mental illness experience is material disadvantage. Material disadvantage involves a loss of resources and opportunities, which can lead to decreased functioning, agency, autonomy, and well-being. When people with mental illness experience discrimination in areas such as employment, housing, education, and healthcare, where they lose out on jobs, apartments, education access, and adequate healthcare treatment, they experience harms of material disadvantage.

One account given of material disadvantage is Kasper Lippert-Rasmussen's (2014a) desert-prioritarian harm-based account of what makes wrongful discrimination wrongful. In his view, discrimination is wrongful when it makes people worse off than if they had not been discriminated against, both through the action at hand and through other discriminatory actions. What the outcome of the discriminatory act should be compared to in making this assessment is not the consequence had the discriminatory act not taken place, but rather the consequence had the discriminatory act not taken place *and* had no other discriminatory acts taken place in the future. This comparison avoids comparing outcomes of different discriminatory acts so the comparison can focus on the specific consequences of the discriminatory act under evaluation (Lippert-Rasmussen 2014a, 158–159).

Lippert-Rasmussen argues that we need to compare how well off people are when they are discriminated against versus when they are not in order to assess whether a discriminatory act harms them and is thus wrongful (Lippert-Rasmussen 2014a, 165–166). In the prioritarian aspect of his account, smaller benefits and harms to those who are worse off have greater effects on those people than larger benefits and harms have on those who are better off. Thus, the worse off are given moral priority in terms of how the harm that befalls them is assessed. Moreover, in the desert component of his account, harms and benefits that affect those who are "deserving" have greater moral value than those that affect people who are "undeserving"; thus, their moral needs take precedence, and the harm they experience matters more, morally speaking. Lippert-Rasmussen does not explain what constitutes deservingness, however, leaving this aspect of his account quite vague, as it is not clear how we should assess who is deserving or not. It is easier to assess who is worse off and who is better off, making the prioritarian aspect of his account easier to understand and apply.

Lippert-Rasmussen does acknowledge other types of harm besides material disadvantage, including status loss and loss of recognition. He points out that some dimensions of a person's experience where harm can occur are "social status, legal recognition, income, education, or freedom from subordination and oppression" (Lippert-Rasmussen 2014a, 161). However, his account does not address the kinds of harms that constitute subordination, oppression, or injustice; for that, we need an account of the power relations involved with discrimination, which I address next.

While Lippert-Rasmussen acknowledges that harm can occur in these different areas, most of his discussion focuses on how "well off" someone is, which I think refers mainly to material advantage or disadvantage. His focus is mainly on how well off someone is, at one point referring to this in terms of resources and welfare (Lippert-Rasmussen 2014a, 103–128). (He addresses a different type of harm, namely that of being demeaned by the discriminatory act, elsewhere and concludes that being demeaned is not what makes wrongful discrimination wrongful. I address this view later in this chapter.) He argues that what matters to whether discrimination is wrongful is the outcome that a discriminatory act has on the person, or what we can understand as the material effects that an act has on a person's welfare or well-being, functioning, and quality of life. This account of discrimination is a strong consequentialist account in which harm involves making a person worse off than they would be without the discriminatory act, which largely has to do with material disadvantage.

He acknowledges that harm can be measured in a multitude of ways and believes his account is compatible with many views (Lippert-Rasmussen 2014a, 160–162). For example, harm can be assessed either in relation to a segment of a life or to one's life trajectory as a whole. Harm can cause an overall effect on a person's life or be multidimensional, affecting people just in a particular area of their life. Harm can be evaluated in relation to a person's welfare or their resources. Moreover, harm can be assessed with regard to its effect on particular individuals versus its effect on a social group as a whole. People with mental illness can experience material disadvantage in many different areas of their life, and evaluating the consequences of discrimination requires assessing the specific losses of resources and opportunities they experience and their effect on a person's total welfare.

Discrimination and Occasional Benefit

One implication of Lippert-Rasmussen's account is that not all discrimination is wrongful. Sometimes discrimination benefits particular individuals or has no effect on them; in these cases, it is not wrongful (Lippert-Rasmussen 2014a, 155–157). In a different paper, he contends that people sometimes benefit from discrimination; in particular, he argues that indirect discrimination can be advantageous globally, for a group on the whole, even if it is disadvantageous locally, to particular individuals; the group as a whole may benefit overall from the indirect discrimination even if particular individuals are harmed by it (Lippert-Rasmussen 2014b). For example, a system that discourages women to enter a certain workforce because they could become pregnant (and pregnancy could interfere with their job responsibilities) hurts particular women who are precluded from this employment field, but if as a result of this system women enter a different workforce that includes better hours, pay, and benefits, then women as a whole benefit from this system even as particular women are harmed by it. As another example, a system that forcibly hospitalizes homeless individuals with suspected mental illness hurts some particular homeless individuals who do not want mental health treatment by violating their autonomy and decreasing what they perceive to be their quality of life; but it benefits homeless individuals with mental illness as a whole by treating their mental illness, which increases their well-being, and consequently enhances their autonomy.

The opposite could also be true, where the group as a whole may be harmed overall from the indirect discrimination, yet individual members of the group may benefit from it. For example, a system of meritocracy could be harmful for Black Americans as a whole, who are usually judged to be inferior to Whites, yet advantageous for particular Black Americans who are regarded as superior. As another example, a system that ignores and discounts mental health patients' concerns is harmful to mental health patients as a whole but could benefit particular individuals whose concerns are delusory and harmful to themselves (so that it is better for them if those concerns are not entertained). Discrimination thus does not always result only in disadvantage, and local and global effects of discriminatory acts must be untangled in order to truly understand the nuance of the effects of this differential treatment.

However, I would caution us to reconsider the ways discrimination can harm people, as even when it appears to benefit people materially, it can still create other types of harm. The fact that discrimination does not always disadvantage people who are discriminated against leads Oscar Horta to question the role disadvantage plays in wrongful discrimination. He argues that either wrongful discrimination does not require disadvantage, in which case what it *does* require must be examined further, or cases where individuals or groups are benefitted by (or at least not harmed by) discrimination do not count properly as wrongful discrimination (Horta 2015). I would counter that even in cases where people appear to benefit from a discriminatory act, namely through material advantages such as attaining resources like housing and healthcare, or having various opportunities such as employment and education, they are still harmed, but in other ways. Even when discriminatory acts materially benefit someone, a person still typically experiences harms to agency and autonomy, harms of degradation, and harms of unfair subordination. People can also be harmed through the perpetuation of prejudice and stereotypes that occur wherever discrimination exists. Wrongful discrimination does not always result in material disadvantage, but it does always result in some form of harm. Counter to Lippert-Rasmussen, there are myriad ways discrimination can harm besides material disadvantage.

Harm as a Comparative Distributive Injustice that Is Not Based on Morally Salient Reasons

Another kind of harm that people with mental illness face when they are discriminated against is when they are prevented from having access to benefits that would otherwise be available. Many of these benefits are the material advantages described in the section Harm as Material

Disadvantage, but here we are conceptualizing the relevant harm as a distributive injustice, an unfair way of distributing resources and opportunities, rather than as a simple consequentialist harm of material disadvantage.

Some cases of discrimination occur as distributive injustices where a person is denied a benefit that they would have access to if they did not have the trait that is the basis for discrimination, or if they are denied a benefit that other people who do not have that trait have access to. Re'em Segev argues that wrongful discrimination occurs when a person is subject to distributive injustice that is not based on a morally significant reason (Segev 2014). Segev understands distributive justice as a fair and equitable distribution of benefits. A distributive injustice involves being denied a benefit when there is no morally compelling reason for the denial. It seems worth adding that another condition of the denial being unjust is when other people have access to the benefit and/or the person would have access to the benefit if not for the morally irrelevant reason that the differential treatment is based on.

A morally significant reason is a reason that justifies differential treatment such as denying someone a benefit that is available to others. For example, qualifications of aptitude that are necessary to carry out a job well often constitute a morally valid reason for why differential treatment of offering qualified candidates a job but not offering this to unqualified candidates is justified. Being denied a benefit for a reason that is not morally significant or relevant constitutes the kind of distributive injustice that Segev thinks characterizes discrimination. (Segev believes that distributive injustice that is not warranted by a morally significant reason characterizes all wrongful discrimination. I agree that this characterizes some forms of discrimination, but not all; some forms of discrimination concern disrespect and treatment of someone as inferior, as well as affronts to agency and autonomy, rather than a distribution of benefits.) It is important to note that the benefit that is being denied through a discriminatory act or policy must be a benefit that is available to other people who do not have the trait that the discrimination is based on and/or would be available to the person who is being denied it if they did not have the trait that is the basis of discrimination.

Being denied housing, education, or employment because a person is known to have mental illness is an example of a distributive injustice that is not justified by a morally significant reason and thus constitutes wrongful discrimination on this account. It is also an example of identity-prejudicial distributive injustice because it is based on prejudicial views about the identity of having mental illness. When a person is rejected as an applicant to rent an apartment because they are known to have a mental illness, or when they are forced to withdraw from school because they have a mental illness, or when they are rejected as a job applicant because they are known to have mental illness, they are being denied a benefit that they would have access to if they did not have the trait that is the basis for discrimination, and that others who do not have this trait have access to. However, having mental illness is not a morally significant reason to justify differential treatment such as denial of a benefit. As a result, housing, education, and employment discrimination of people with mental illness constitutes wrongful discrimination in part because it involves being denied a benefit without a morally significant reason for the denial.[1]

Harm as a Baseline Distributive Injustice that Prevents Equal Participation in Society

Another type of harm people with mental illness can experience due to discrimination is being hindered in being able to participate equally in society by being denied the basic goods that are required for such participation. In the previous section, I described Re'em Segev's account of discrimination as a distributive injustice not justified by a morally significant reason, where the distributive injustice consists of being denied benefits that others have access to, and that the person being discriminated against would have access to if they did not have the trait that is the basis for discrimination. Here I want to explore a different kind of distributive injustice that people can experience as a result of discrimination: being denied basic goods that are necessary for being able to participate in the social, moral, epistemic, political, and economic practices that other members of a community participate in.

[1] There are contexts where it could be argued that it makes sense to treat people with mental illness differently than others. For example, when a person with mental illness is in imminent danger of harming themselves and/or others, they are legally and perhaps morally justified in being put on a 72-hour mental health hold where they are involuntarily committed to a psychiatric hospital or unit. However, in this example, it is the imminent danger of harm that justifies the involuntary commitment, not the condition of having mental illness. When a person with mental illness is out of control, dangerous, and/or violent, they may be justifiably forcibly injected with medication such as antipsychotics that calm them down. Again, however, it is the condition of being out of control, dangerous, and/or violent that justifies the forcible injection, not the condition of having mental illness. When people with mental illness are justifiably treated differently than others, the morally salient reason that justifies the differential treatment is their behavior, and the risk of harm their behavior presents, not the condition of having mental illness.

Sophia Moreau characterizes the relevant denial of benefits involved with the distributive injustice aspect of discrimination as a loss of access to basic goods (Moreau 2020, 124–151). For her, the loss that is relevant is not of benefits in general that can advantage people in various ways, but more specifically a loss of basic goods that are a condition of basic functioning and well-being. Basic goods are what people need in order to be able to participate, and to be *seen* as being able to participate, equally in society compared to others (Moreau 2020, 125–126). Treating people equally, in Moreau's view, involves enabling access to the basic goods necessary for participation in various social, moral, epistemic, political, and economic practices inherent to living in society. Precluding people's access to these goods prevents them from being able to participate equally in the practices that make up society.

Thus, what is distributed unjustly for her are the resources and opportunities needed to have one's basic needs met in order to achieve a level of basic functioning and well-being. This is a narrower view of what goods are morally relevant in assessing fair distribution compared to Segev's broader view of benefits in general that can advance people's functioning and well-being in various ways. Equal access to the same resources and opportunities that others have, whatever those are, and whatever level of functioning and well-being a person is at, is important for fair distribution of resources and opportunities, and thus is an important aspect of distributive justice. However, another important aspect of distributive justice is meeting or exceeding a minimal threshold of what people need for adequate functioning and well-being, here understood as being able to participate equally in society. Being denied access to basic goods needed for equal participation is another form of distributive injustice that wrongful discrimination can consist of.[2]

Basic goods include all resources and opportunities that are necessary for participating equally in society. These include provisions like nutritious food, safe drinking water, adequate shelter, and suitable clothing, as well as resources and opportunities including basic healthcare, rudimentary education, employment, and sufficient income to be able to purchase what one needs. Moreau argues that what constitutes a "basic" good is relational and contextual: What a person needs in order to be able to be full and equal

[2] One of the basic goods that people need in order to be able to participate equally in society is basic healthcare. Norman Daniels argues that people have a right to healthcare as one of the "positive" welfare rights that is a requirement to meet people's basic needs and enable equality of opportunity. Mental healthcare, as one form of healthcare, would presumably be one of the basic goods people need to get met (Daniels 2009, 364–366; Daniels 2008, 313–332).

participants in their society depends on the nature of their society and on their individual relationship to their community and government (Moreau 2020, 134–135). However, even if we reject a contextual understanding of what constitutes basic goods, we can determine a set of basic goods that everyone needs regardless of their society, and these include various basic goods, which are also recognized in the United Nations International Covenant on Economic, Social and Cultural Rights (United Nations 1966).

For Moreau, there is no morally salient reason that can justify denying people access to basic goods. Any denial of access, regardless of the reason, damages a person's functioning and well-being and prevents them from being able to participate as an equal member of society. Having access to basic goods is a matter of treating people equally, not a matter of comparable distributive justice, where what is relevant is what resources and opportunities a person has access to *relative to* other people. While Segev's account of wrongful discrimination as a distributive injustice not justified by a morally significant reason – a denial of benefits that others have access to, and that one would have access to if not for having the trait that is the basis of the denial – Moreau's account of wrongful discrimination consists of a different form of distributive injustice. It is not an injustice because of the way one's own situation compares to that of others, but rather because one's own situation does not meet basic requirements for being able to participate equally in society. In other words, it is a threshold conception of distributive injustice rather than a comparative conception.

When people with mental illness are denied through discrimination basic goods essential for their survival, such as housing, employment of some kind, and basic healthcare, they experience a harm that consists of being a distributive injustice preventing equal participation in society. People with mental illness who are also homeless experience this type of harm when they do not have regular access to safe and stable shelter. Homeless individuals are prevented from participating in many facets of society due to their homeless status and are often not recognized or treated by others as equal participants. In fact, they are regularly shunned from many areas of society. People with mental illness who lack access to basic healthcare or who lack access to regular meals are also prevented from partaking in certain areas of society when they are too sick or too hungry to be able to participate. Discrimination in healthcare, education, employment, and housing can cause this lack of access.

Lacking basic goods is a major problem for people with mental illness because their illness often makes them too poor to be able to seek necessary provisions themselves, and this poverty sometimes results from discrimination in areas such as employment. People with mental illness are often viewed as being incapable of carrying out the tasks and responsibilities that are associated with various jobs and are thus denied employment because of assumptions about their lack of competence or propensity for violence. When people are unemployed, they are unable to buy or access many of the basic goods they need. Thus, many people with mental illness experience the harm of distributive injustice that precludes them from taking part in society.

Harm as Unfair Subordination

Another kind of harm that people with mental illness can face when they are discriminated against is structural injustice that occurs as a result of unjust power relations. Sophia Moreau says that many acts of discrimination constitute unfair subordination, where social structures are arranged in such a way that some people are subordinated to others and have less power and agency over their situation and their place in society as a result (Moreau 2020, 39–75). Unfair subordination is a form of inequality that subjects certain people to more constraints on their action by virtue of the way they are socially situated relative to others. It occurs when people are subordinated on account of having some trait that is devalued in society, so that they are seen as having lesser worth on account of possessing that trait.

Moreau defines social subordination as a relationship between different social groups, where some have greater status, including moral status, than others. This disparity in moral status is *in relation to* others, where those with greater status possess their status in part through the subjugation of those with lesser status. It occurs when "one social group has a standing in society as a whole that is lower than that of another social group" (Moreau 2020, 50). A social group is a collection of members who share a certain trait, where that trait is treated as socially salient, "in the sense that others in society take that trait to have implications for the character and behavior of members of that group, and for the social roles that they are capable of occupying" (Moreau 2020, 50–51). The trait is picked out as a salient feature of a person's identity or behavior that marks them as being a certain type of person who occupies certain kinds of social roles. People who have devalued traits are treated as lesser persons or moral agents and are seen to occupy very limited social roles, which tend to be viewed negatively. On the other hand, people who have valued traits are treated as having more worth and more agency and are accepted as occupying a greater range of social roles, where most of the social roles are deemed positive.

People who have devalued traits are positioned in society as having less power and authority over their lives compared to people who lack these traits and who are thus considered "normal." They are positioned within social structures in ways that create greater constraints on their action compared to others, giving them less power and agency (Young 2011, 43–74). This subjects them to structural injustices of being unfairly and disproportionately burdened by constraints and limitations caused by social structures. In contrast, people who are seen as having greater worth and who are socially positioned in ways that allow social structures to grant them unfair and disproportionate benefits are able to have more power, agency, and authority. One form of power and authority that those who have greater worth possess is epistemic power and authority.

Moreau understands power as having power over someone, while authority (or what she calls "de facto authority") involves having the power to encourage others to follow them, where the others do not feel compelled or coerced but rather believe they are acting on their own free will (Moreau 2020, 52). Social structures are designed and set up in such a way that they enable some people to have de facto authority over others, where those with greater authority can more easily convince others to follow their recommendations or requests in a way that leaves those others believing they have done so of their own volition. Moreau says that de facto authority involves "the power *to be listened to*, and to be taken seriously when one brings a complaint against another" (Moreau 2020, 52, italics original). This authority involves being seen as having the credibility to be listened to and have one's needs and interests taken seriously and responded to, and to not be ignored, discounted, or dismissed. De facto authority thus generates the epistemic power to be treated as a credible and legitimate epistemic agent capable of making credible and trustworthy epistemic contributions. Those without this authority lack this epistemic power. Insofar as this epistemic power stems from how a person is situated in society in relation to various social structures, the inequality of epistemic power that people who are viewed as having greater worth have compared to people who are regarded as having lesser worth constitutes structural epistemic injustice (Anderson 2012).

Part of what upholds unjust subordination is structural accommodations, which are the way structures in the social or physical environment are designed to respond to the needs and interests of some people – those with more power, privilege, authority, and credibility – while ignoring or discounting the needs and interests of others – those who are marginalized in society and who are seen as lacking authority and credibility. Structural accommodations are *structural* in that they consist of features in the social or physical environment, not attitudes, actions, or dispositions of behavior of individuals. Structural accommodations can also be understood as structural *privileges* that prioritize the needs and interests of some over others. Where structural accommodations exist, the beliefs, knowledge, values, desires, actions, needs, and interests of those with power, privilege, authority, and credibility are seen as normal and treated as the norm, while the beliefs, knowledge, values, desires, actions, needs, and interests of marginalized groups are seen as exceptional and often treated as problematic (Moreau 2020, 43). With this seeming normalcy, structural accommodations tacitly accommodate the needs and interests of more privileged groups over the needs and interests of marginalized groups (Moreau 2020, 56).

In addition to including structural accommodations, unfair subordination also consists in part of being granted different levels of consideration, or deference and respect. Moreau references Niko Kolodny as arguing that those with greater power and authority are seen as having valued traits that "attract positive responses of deference and respect" (Moreau 2020, 53; see Kolodny 2014). With this greater consideration, the needs and interests of people who have greater power and authority are given more priority and are often viewed as normal and normalizing. What *they* believe, know, value, and desire is viewed as the *correct* beliefs, knowledge, values, and desires to have and pursue, and these are given uptake by others in that they are given priority in being addressed, responded to, and built upon.

A lack of consideration can consist of censure. Censure occurs when those with a devalued trait are seen as lacking worth, so their beliefs, knowledge, values, desires, and actions are also seen as lacking value (Moreau 2020, 53). Their beliefs, knowledge, values, desires, and behavior are censured in that they are not given consideration. They are ignored, discounted, or dismissed, not taken seriously, and not given uptake by being addressed, responded to, and built upon. People who have valued traits garner greater consideration while people who have devalued traits attract censure (Moreau 2020, 54).

Unfair subordination consists of the injustice of society being organized so as to treat people who are seen as lesser in ways that disregard their needs and interests, discount their views, and censure them. When people are subordinated, social structures exist that support their subordination, so they are treated as lesser in what power, privilege, credibility, and authority they are granted (sometimes virtually none). Unfair subordination is not just about how a person is regarded by others but also about what power, privileges, authority, and credibility they are allowed to have given the rules, limitations, and opportunities set up by society. Many cases of discrimination unfairly subordinate people by structuring society in ways that create more obstacles, restrict opportunities, decrease avenues for agency, reduce privileges, and undermine authority and credibility.

People who are unfairly subordinated in one way are often unfairly subordinated in multiple ways, because they often have multiple social identities that give them less power and privilege in society. Having less power and privilege in one area of one's life tends to perpetuate a lack of power and privilege that applies to other areas of one's life. For example, a person who is homeless is unfairly subordinated on account of their homelessness, but being homeless is, for many people, connected to having physical or psychiatric disabilities and/or minoritized race that also subordinate a person. The way lack of power and privilege in one area reinforces it in other areas as well can subjugate a person so much that it can be hard to see how they can find their way out of that web of powerlessness.

People who have mental illness often experience unfair subordination as a result of being discriminated against. As epistemic justice theorists point out, they are afforded less credibility and authority than others due to their mental illness and the negative stereotypes such as incompetence, neediness, dependence, and defectiveness that are associated with mental illness (Kurs and Grinshpoon 2018; Sanati and Kyratsous 2015). As a result, they are often censured, and their needs, ideas, and contributions are not given adequate consideration. Moreover, many structures in society are hard for people with mental illness to navigate because they are not set up to accommodate people who have difficulties with reasoning, emotional regulation, volitional skills, attention, social skills, and communication skills. Social structures are typically designed to help people who are able to express themselves clearly, control their emotions, identify their needs, and have ideas of their own. People who are unable to do these things often cannot take advantage of the benefits that different social systems can bestow and so they lose out. This is a problem not of the individuals who lack relevant skills and capacities but rather of the social structures that are designed to accommodate only a narrow range of mental processes and behavior. Many of the harms that people with mental illness face result from the unfair subordination involved with how social structures are organized to advantage those with greater power and privilege and ignore the needs, ideas, and contributions of those with less power and privilege.

Harm as a Violation of Dignity: The Harm of Being Demeaned

Another way that people with mental illness can be harmed through discrimination as well as microaggressions is by being demeaned through differential treatment. When a person is demeaned, they are treated as being unequal to others in the sense that they are seen as having less moral status and being less of an agent, person, or human being. Being treated as unequal and lesser compared to others violates a person's dignity. Deborah Hellman argues that a "bedrock moral principle" is that all people have equal moral worth (Hellman 2008, 6). She identifies two sub-principles that accompany this primary moral principle. First, since everyone has inherent moral worth, intrinsic value, and dignity, all persons are obligated to treat other persons with respect. Second, other traits, including social traits like gender, race, class, ability status, nationality, and sexual orientation, do not affect the inherent moral worth, intrinsic value, and dignity of persons. Treating someone as being unequal and lesser compared to others is a violation of dignity and moral worth; it is a way of treating someone as having lesser (or no) intrinsic value, or in Kantian language, it is a way of treating someone as not being an end in themselves. It can also be understood as a way of treating someone as subhuman, dehumanizing them by making them seem to be less than fully human, as discussed in Chapter 5.

Hellman identifies this treatment of others as unequal and lesser "demeaning." She says:

> To demean is to treat someone in a way that denies her equal moral worth and thus picks out a wrong that is intimately tied to the value that underlies our worries about differentiation in the first place. To demean is not merely to insult but also to put down, to diminish and denigrate. It is to treat another as lesser. (Hellman 2008, 29)

Lesser treatment can consist of many different types of acts, but what they all have in common is that they convey a message to the person receiving the treatment that they have less intrinsic value and less moral worth than others. As discussed in Chapter 5, stigmatizing a group of people based on a social trait they share leads to status loss and dehumanization, viewing others as lesser and as less than human. Discrimination and microaggressions point to that status loss and tell the person who loses their status as a full human being that they have insufficient worth.

Demeaning others is only possible in the context of power relations in which some people are socially situated in a way that they have greater power and others are socially situated in a way that they have less. Those with greater power have the power to affect the outcomes of those with less power, so that they can exert control over what happens to those with less power. A person can only demean others when they have power over others such that their act subjugates and denigrates them (Hellman 2008, 35). Demeaning others is an act of power in which one's debasement of another is sanctioned and supported by one's more privileged social position.

Hellman argues that what makes wrongful discrimination and microaggressions wrongful is the way that they demean people by putting them down. Discrimination and microaggressions consist of words or actions that "tell" the person that they are worth less than others. Sophia Moreau characterizes this account of discrimination as an "expressivist" view where what matters to discrimination is that it conveys a message to people who are discriminated against, namely the message that they are worth less than others (Moreau 2020, 45). It is unclear whether Hellman herself sees her view as primarily expressivist, or if she thinks demeaning treatment includes acts of treatment that go beyond conveying messages.

Hellman notes that actions have meaning given by the cultural and historical contexts in which these actions take place (Hellman 2008, 25). Actions that are of a type that have historically been viewed as treating some people as lesser compared to others, or that occur within a cultural context that can be interpreted as treating some people as lesser compared to others, take on that meaning of degradation even in new scenarios. An act such as preventing a person with mental illness from renting an apartment does not stand in isolation; it takes its meaning from a cultural context in which many people with mental illness are denied living quarters based on stereotypes about what it is to have mental illness. This act is a token of a type of action that occurs commonly within American twenty-first-century culture; thus, this act must be interpreted in light of this context.[3]

Demeaning acts occur within a context of power, where people who are capable of demeaning others have power over people who are capable of being demeaned (Hellman 2008, 35). This power, too, occurs within a historical and cultural context where some groups of people have or have had power over what happens to other groups of people. When a person who is a member of a social group that has historically been

[3] Raino Malnes objects to the notion of discrimination as always taking its meaning from other similar acts (tokens) that share the same type, arguing that discrimination can occur whether the discriminatory act is performed across various social contexts, or whether it is an idiosyncratic act that just occurs in a specific context (Malnes 2015). Hellman does not claim that discrimination *always* takes it meaning from similar acts, only that it often can.

treated as lesser, or is a member of a social group that is currently systematically disadvantaged, and they are treated with disdain through an action that conveys disrespect, they are demeaned and degraded as members of that social group. Their denigration takes place and gains its meaning from cultural and historical contexts where similar actions by some groups that have greater power have denigrated other groups of people who have less power.

Moreover, when a person is a member of multiple social groups that have been treated as lesser or is currently systematically disadvantaged, they are disdained in a plurality of ways that reinforce each other. A person who is Black, or otherwise is a person of color, has physical disabilities, has mental illness, is homeless, and/or uses drugs can be demeaned on account of all of these aspects of their identity, and the disrespect they receive in one area can reinforce the disrespect they receive in another area, making them multiply demeaned.

When individuals with mental illness are demeaned, the act of discrimination or microaggression that demeans them occurs in a context where people with mental illness have historically been treated badly and discriminated against, and in a cultural milieu where negative stereotypes about mental illness that can be used to justify treating people with mental illness poorly persist and pervade the culture. A single act of demeaning someone with mental illness signifies a cultural and historical practice of demeaning all or many people in that social group, which gives more meaning to the solitary act and has more force on the person subject to the act, affecting them more deeply. When discrimination demeans people by pointing to their lesser status, it often carries weighty meaning that has greater impact on a person than the act alone would seem to have. Demeaning people is a significant harm that many individuals with mental illness experience when they are discriminated against.

Harmful Agential Effects: Loss of Deliberative Freedom

One of the harmful effects that wrongful discrimination and microaggressions have on people with mental illness is the loss of deliberative freedom. This loss of freedom can be understood as one of the costs involved with discrimination and microaggressions that are imposed on the person subject to these. Paul De Font-Reaulx identifies this feature of discrimination – its costs – as what constitutes the wrongfulness of discrimination. Wrongful discrimination creates harmful externalities for individuals who have the trait that is the basis of discrimination, where externalities are understood as costs or benefits accruing to a person as a result of discrimination (De Font-Reaulx 2017, 110). These harmful externalities include having to pay constant attention to the socially salient trait that is the basis for discrimination and microaggressions, and it causes a loss of deliberative freedom.

Sophia Moreau argues that the costs of how people are treated – whether this is equally or differently (and sometimes with discrimination) – must be borne by someone. The ethical question at the heart of discrimination is "Who should bear the costs of how a person is treated, whether the person is treated equally (fairly) or unequally (discriminated against)?" (Moreau 2020, 197–198). Moreau notes that the question of who should bear costs rests on a thin notion of responsibility that involves responsibility for cost. The cost of being discriminated against or subject to microaggressions is usually borne by the person who is targeted, because they have to regard a devalued trait they possess as salient even in contexts where it should be irrelevant. This shows up in the expressive harm discussed in Chapter 3, where stereotypes are always in the atmosphere and must be acknowledged and responded to even when they are repudiated, so that individuals always have to pay attention to their thoughts, feelings, and behaviors because of how their trait makes them appear to others. This also shows up in attendant stigma, where individuals are always aware of the possibility of being judged by others on account of possessing the stereotyped trait. Discrimination and microaggressions, too, always keep individuals aware of their thoughts, feelings, behavior, and identity in relation to the social trait that others find socially salient.

This cost is a loss of deliberative freedom, which Moreau says is one of the aspects of wrongful discrimination that makes it wrongful. Being discriminated against or subject to microaggressions is very costly, requiring a degree of moral energy and effort that people who are not a target of these do not have to experience. Thus, it creates an undue burden on the person. This burden interferes with deliberative freedom, a freedom that Moreau argues people have a right to (Moreau 2020, 77). Deliberative freedom is the freedom of thought and action that enables careful reflection and choice, and consequently allows one to be able to shape the course of one's life in the way one wants through one's deliberations, choices, and actions (Moreau 2020, 81). In other words, deliberative freedom is the freedom to think carefully about choices and actions in a way that enables one to have and express autonomous agency.

Discrimination and microaggressions interfere with deliberative freedom by making the possession of a certain social trait an issue in a context where it shouldn't be relevant (Moreau 2020, 83). As discrimination and microaggressions involve singling out people based on a devalued trait they possess, it brings that trait to the forefront of thought and action, exposing the possession of the trait and marking the person who has the trait as *someone who has that trait.* When discrimination or microaggressions do not occur, the possession of that trait is in the background, something that exists but is not noticed or paid attention to because it is seen as something that is not salient to thought and action. When discrimination or microaggressions do occur, on the other hand, the trait gets foregrounded and is noticed and paid attention to, even in contexts where it seems the trait should be irrelevant. The trait is made salient through that attention, and its salience shapes thought and action.

Based on the way discrimination and microaggressions make a trait salient even in contexts where it should be irrelevant, Moreau describes deliberative freedom as the freedom to deliberate about one's life and make choices without having to treat such a trait as salient and give the trait undue attention, as well as without having to bear the costs associated with having the trait. Deliberative freedom involves the freedom to not have to think about the trait as something worth paying attention to and to not have to do anything about the trait; it is the freedom to treat the trait as a background condition that does not have to be given salience, and the freedom to not have to bear the burden of the cost associated with the trait. She says: "Deliberative freedom is the freedom to deliberate about one's life, and to decide what to do in light of those deliberations, without having to treat certain personal traits (or other people's assumptions about them) as costs, and without having to live one's life with these traits always before one's eyes" (Moreau 2020, 84).

Deliberative freedom involves freedom from both fixed and opportunity costs associated with granting a trait salience, as discrimination and microaggressions impose both fixed and opportunity costs on a person. Having to think about a certain trait by always having it foregrounded as something one must pay attention to and do something about in social interactions is a fixed cost because it is a cost that always exists when the trait is made salient (Moreau 2020, 85). Discrimination and microaggressions also impose opportunity costs that can make it more expensive or difficult to pursue certain options over others. This has the effect of either requiring a person to undergo more hardship when pursuing a desirable but expensive option or of preventing a person who cannot afford to pay the cost from pursuing that option. In this way, discrimination and microaggressions narrow options by making some too difficult to pursue.

Privilege is often self-obscuring: it often makes invisible to the people who have privilege what privilege they have. The privilege of having deliberative freedom, therefore – the freedom to not have to notice, pay attention to, and act in relation to having a certain trait – often blinds people from noticing the deliberative freedom they have (Moreau 2020, 87). Only when people lack this freedom do they tend to notice that it is a freedom they should have and be aware that they do not have it. Privilege thus imposes a power to not have to be aware of what one is privileged in possessing on those who possess it. Moreau says that privilege "is a power not to be bound by or burdened by certain assumptions and certain costs" (Moreau 2020, 87).

Of course, deliberations and choices often involve costs, sometimes high costs that are prohibitive and result in closing off certain options, and these are not always unfair; such costs do not always result from discrimination. Deliberations can be constrained legitimately by acts and choices of others (Moreau 2020, 88). To determine whether costs and constraints are fair or not, we need to examine whether they result from personal choices one has made or whether they result from a person being singled out for having a devalued trait, as in discrimination or microaggressions. Moreau says we need to examine "whether they [costs and constraints] reflect other people's assumptions about who [a person] is and what roles [they] ought to occupy" (Moreau 2020, 90).

Whether these costs and constraints are unfair also depends on whether they "affect goals or choices that are particularly important to the discriminatee's own conception of herself and of her life; and whether most people in the discriminatee's society, too, face the kind of deliberative burden that she is facing" (Moreau 2020, 90). If the person faces these costs and constraints due to possessing a devalued trait, this affects the person's self-concept and self-understanding in a way that other people who do not possess the trait do not have to experience. When a person is singled out as having a devalued trait that is viewed as salient in a context where it should not be relevant, they are exposed to costs and constraints that others, who do not have this devalued trait, do not, and this creates an undue burden on the person. Moreau concludes that constraints are wrongful when they are based on other people's assumptions and stereotypes about a person based on a devalued trait they possess (Moreau 2020, 94).

People with mental illness often experience this loss of deliberative freedom when they are discriminated against or the target of microaggressions, because the discrimination or microaggression makes them pay attention to their mental illness, and the thoughts, feelings, and behavior they exhibit in relation to their mental illness, in a way that those who are not discriminated against do not have to do. Some people with mental illness do not pay attention to this because they may lack insight, or the awareness that they have mental illness; or they may simply lack self-awareness about how they are regarded by others so they do not consciously think about how they think, feel, and act in relation to how they come across to others. However, many people with mental illness do have insight and self-awareness and thus do experience the harm of having to pay attention to their illness in contexts where it should not be relevant. This results in a loss of deliberative freedom for them, which damages their agency.

Harmful Agential Effects: Impingements on Autonomy

People with mental illness also experience impairments to their autonomy when they are discriminated against and to a lesser extent when they are subject to microaggressions. Discrimination impinges on autonomy in a few ways. Notably, insofar as discrimination is a distributive injustice, it decreases the opportunities a person has access to and thus the options for action that they could choose from. In denying people a benefit that they would otherwise have available to them, and in preventing them from having access to basic goods, discrimination reduces what resources and opportunities are available to a person, which closes off options. This impacts both self-determination and self-governance, two important aspects of autonomy.

One condition of the realization of autonomy is having multiple good options to choose from (Raz 1986, 204; Stoljar 2014). This is an opportunity condition of the self-determination aspect of autonomy in that meaningful choice requires meaningful options to choose from in order to have greater opportunities in one's social environment (Mackenzie 2014, 17). It is also a condition of the self-governance aspect of autonomy: Having multiple good options both enhances competence by enabling stronger reasoning for one's choice and increases voluntariness by enabling a choice to be less constrained by one's circumstances. Consequently, having multiple good options enables a person to have more control over what happens to them in more areas of their life and also enables their choice to be more their own. This allows them to have more power in directing their life as they choose.

Having multiple good options is enabled by positive structural conditions that increase the opportunities available to a person, while negative structural conditions reduce these. Negative structural conditions such as discrimination, microaggressions, oppression, marginalization, economic scarcity, and violence reduce what opportunities a person has access to by making some desirable options unavailable to a person. Positive structural conditions, on the other hand, such as not being subject to injustice, being socially included, being afforded respect, having economic power, and not being fearful of violence or assault enable greater opportunities by making many desirable options possible (Sherwin 2000, 79). Discrimination and microaggressions limit options by preventing people from having the same opportunities for accessing resources and participating in activities that other people have. In the context of mental health, mental illness stigma and discrimination frequently prevent people from accessing safe and secure housing and getting and holding down a job (Corrigan and Kleinlein 2005, 18–20; Rüsch, Angermeyer, and Corrigan 2005, 530; Stuart 2006, 523; Wahl 2012, 9–10). As discussed in Chapter 1, people with mental illness report being discriminated against in many areas of their life, including housing, employment, education, and close relationships (Angell, Cooke, and Kovac, 2005, 73–75; Farrelly et al. 2014; Thornicroft et al. 2022). Such discrimination decreases what opportunities are available to a person by closing off what would otherwise be desirable options and preventing people from being able to access these. Moreover, microaggressions can cause a person to refrain from participating in various activities and accessing resources and opportunities out of fear of disrespect, invalidation, or deeper discriminatory acts. This also closes off desirable options.

In addition to decreasing opportunities for making various kinds of choices, having reduced options also affects the competence and voluntariness conditions of the self-governance aspect of autonomy. Being able to make a meaningful choice involves being able to give strong reasons for one's choice, which often requires having multiple good options to choose from. Having a sufficient range of suitable options gives people a variety of possibilities they can choose from, which allows them to consider alternate reasons for action. When they have multiple good options, a person can consider what the outcome of different variables would be and compare different outcomes and various courses of action. With the ability to compare different options, a person can consider an array of disparate reasons for action and decide which seem weightier and more urgent. Having multiple good options thus enhances the reasoning process involved with choice, increasing competence. When people have severely reduced options, as they do when they are discriminated against or subject to microaggressions, they are unable to consider different outcomes and various courses of action and able to give much more limited reasons for their choice. This decreases their competence around that choice.

Having reduced options also decreases voluntariness or authenticity. When a person feels like they have only one option or very few options, they typically feel more pressure to choose that option and may feel like they make no choice at all.[4] In such a case, they are unduly influenced by the constraining circumstances around that choice: The fact that there are no or few other options makes them feel forced to choose the one option available. When this happens, a person lacks voluntariness or authenticity in their choice.

A choice is voluntary and authentic when it comes from oneself and not from external factors (Feinberg 1986, 104; Feinberg 1983, 7). One test of voluntariness and authenticity is reflective endorsement: If a person is capable of evaluating their choice and giving reasons for their choice that are intelligible to others and that they can endorse as reasons they themselves believe in, this enables them to endorse their choice as their own as a result of this evaluation (Dworkin 1988, 20; Frankfurt 2003; Watson 2003). When a person cannot reflect on and evaluate their choice, they are unable to endorse it and thus unable to own their choice as theirs. The fewer options a person has, the harder it is to evaluate a choice properly because it is harder to give reasons for it that come from within, rather than from the pressure of having few options and feeling coerced into choosing the only option(s) available. Having multiple good options is thus important not just for self-determination but also for self-governance.

In reducing options, discrimination and microaggressions do not just limit what opportunities are available to a person; they also decrease the competence and voluntariness of a person's choice. In allowing people to have fewer options, discrimination and microaggressions create coercive circumstances where a person has to choose the least bad option from a set of bad options, decreasing the voluntariness of a choice. They limit what reasons a person has for making their choice so their choice is less rational.

[4] Aristotle calls a choice between two bad options a mixed choice, in that there is a sense in which the choice is voluntary – the person is making a choice – and a sense in which it is not – the person would not choose either option if they had a better alternative. A mixed choice between two bad options can feel like one is making no choice at all. Thus, while the choice may be partly voluntary, it does not *feel* voluntary. With such a choice, the reasons that a person can give for their choice are much more limited and seem to come from the circumstances surrounding the choice rather than from the person making the choice. This drastically decreases the voluntariness of the choice (Aristotle 2011, 42–43 (lines 1110a4-1110a23)).

In these ways, stigma, discrimination, and microaggressions undermine self-governance.

When people with mental illness are discriminated against or subject to microaggressions, they experience these losses of autonomy. Discrimination restricts their options, for example by limiting what types of housing, jobs, and social opportunities are available to them. This decreases their opportunities for making various kinds of choices, reducing their capacity for self-determination. Microaggressions reduce options by discouraging those targeted from participating in activities and accessing resources and opportunities. In restricting their options, discrimination and microaggressions also decrease the competence and voluntariness conditions of the self-governance aspect of autonomy. Having fewer options compounds the diminished competence they may already have due to their illness, interfering with the capacities and skills required for valuing, making choices, and directing their lives. Having fewer options also makes it harder for a person to evaluate their choice properly and be able to reflexively endorse it as their own. In many ways, discrimination and microaggressions damage autonomy.

Conclusion

When people with mental illness are stigmatized by others, this often takes the form of discrimination, or treating people with mental illness in different, unequal, and worse ways compared to other people on account of their having mental illness. This can also take the form of microaggressions, which are subtle speech and behavior that devalues, discredits, disrespects, and demeans a person. Discrimination and microaggressions based on having a socially salient social trait such as mental illness are wrongful because, in treating people who have that trait unequally and worse than others, they cause many types of harms to those people. As this chapter shows, people with mental illness experience many of these harms. These include material disadvantage of lacking resources and opportunities that are important for functioning and well-being; distributive injustice that prevents a person from having access to the same resources and opportunities that others have access to (or that they would have access to if they were not discriminated against), for no morally salient reason; and distributive injustice that precludes people from having access to basic goods required for equal participation in society. Discrimination also often unfairly subordinates people with mental illness through structural accommodations that discount their needs and interests and ignore their ideas and contributions, censuring them and failing to give them consideration.

When people are denied housing, employment, adequate healthcare, and education access by virtue of discrimination, they typically experience all of these harms.

Other harms occur as well. Discrimination and microaggressions demean people with mental illness by degrading and devaluing them, sending the message to them that they have less moral worth compared to others. Discrimination and microaggressions also impair the agency of people with mental illness by making them have to pay attention to their illness in a way that others do not have to do, diminishing their deliberative freedom. In addition, discrimination and microaggressions impede the autonomy of people with mental illness by restricting their options, which limits the opportunities people have and the choices they can make, and which decrease the competence and voluntariness involved with making choices. In harming people with mental illness in all of these ways, discrimination and microaggressions, which treat people with mental illness differently, unequally, and worse than other people, are wrongful.

So far in this book, I have explained the many harms caused by mental illness stigma, including both public stigma imposed on people with mental illness by others and internalized stigma that people with mental illness project onto themselves. The following chapter shifts gears to examine some of the ways both public stigma and internalized stigma can and ought to be addressed. Reducing mental illness stigma is paramount in order to increase the functioning, well-being, agency, and autonomy of people with mental illness, and to improve their self-concept, self-worth, and other self-regarding traits.

Mental illness stigma is pervasive, in part because negative stereotypes about mental illness persist in the cultural milieu regardless of attempts to repudiate these, and in part because many people simply endorse or perpetuate negative stereotypes about certain groups of people with mental illness, including those with severe illnesses like schizophrenia, those who are homeless, those from racially minoritized groups, and those who have physical disabilities. Self-stigma, attendant stigma, prejudice, and discrimination are widespread as a result. However, mental illness stigma can and ought to be addressed in order to decrease these so that people with mental illness can live better lives that are free of stigma and its damaging effects.

CHAPTER 7

Interventions to Address Stigma

Introduction

Mental illness stigma is far-reaching, pervasive, pernicious, and exceedingly harmful in all its forms. Whether people impose stigma on themselves through internalized stigma or whether they are subjected to stigma through public stigma that can take the form of stereotyping, prejudice, implicit bias, microaggressions, discrimination, epistemic injustice, moral distancing, belonging uncertainty, and/or social exclusion, they experience myriad harms as a result of this stigma. These harms create a moral imperative to address mental illness stigma by diminishing and ultimately eradicating it. Reducing and eventually eliminating mental illness stigma is necessary for people with mental illness to live better lives and to flourish and thrive within the community.

Many different types of interventions to address stigma have been proposed. Some suggestions have been studied for their effectiveness in the field of social psychology. The empirical evidence for the effectiveness of these proposals is reviewed here. Other interventions, especially those that involve philosophical discernment, have not been researched explicitly in social psychology, but they emerge from understanding the ways stigmatization works and the ways stigma causes harm to people who are stigmatized. These are developed here as well.

Interventions to challenge stigma include individual actions that are performed by the person who is stigmatized themselves, individual actions undertaken by other people who have the propensity or possibility of stigmatizing people with mental illness, structural changes that diminish or eliminate structural stigma and improve the lives of people with mental illness, and social and cultural changes that involve affirming the belongingness of people with mental illness and transforming social norms to be healthier and more supportive of vulnerable people. Let us examine some of the interventions that fall under each of these categories.

Individual Interventions for People with Mental Illness to Address Internalized Stigma and Cope with Public Stigma

Individual responses to stigma undertaken by the person who is stigmatized include efforts to help individuals cope with stigma, prejudice, and discrimination; acts of resistance to challenge public stigma imposed upon them by others; and measures that help them resist self-stigma and other forms of internalized stigma. These interventions enable a person to not let stigma control their lives. Challenging both public stigma and self-stigma can involve psychological interventions, philosophical interventions, belongingness, support, self-disclosure, and advocacy work.

Psychological Interventions

One psychological intervention is to develop mindfulness, which increases positive reappraisal, decreases rumination, alleviates negative emotions, and reduces stigma stress (Aliche et al. 2024; Fung, Liu, and Wong 2022; Wang et al. 2023). Mindfulness consists of being in the present moment, and it is a way of resisting negative emotions like anxiety and fear that are involved with internalizing stigma and reducing the stigma stress that comes from facing public stigma. Stigma stress is the stress involved with seeing stigma-related harm as excessively damaging and believing that dealing with it requires effort that exceeds one's coping capacity (Wang et al. 2023, 808). Mindfulness helps a person avoid stigma stress so that stigma appears to be something that they are capable of dealing with, enabling self-efficacy.

In addition, mindfulness reduces rumination and enables positive reappraisal, which involves reassessing an event in a positive light (Fung, Liu, and Wong 2022). With positive reappraisal, a person can assess an event they formerly considered to be a threat, challenge, or harm in a different context to see either that it has positive aspects or that it can be dealt with (Aliche et al. 2024, 3125). Often, for people with mental illness, threats, challenges, and harms can feel insurmountable and cause distress and despair over a person's seeming inability to deal with them. With positive reappraisal, a person can recognize that they have the capacity to deal with the adversity in a way that allows them to move on with their life and not feel trapped by the adversity (Aliche et al. 2024; Wang et al. 2023). Moreover, mindfulness helps a person avoid ruminating over the thoughts and feelings that cause them stress, anxiety, and depression, so these no longer rule their life, enabling a person to better resist internalizing stigma and helping a person feel like they can cope with public stigma. By increasing positive reappraisal and reducing rumination, mindfulness decreases stigma stress, reduces the internalization of stigma, and increases well-being (Wang et al. 2023).

Another psychological intervention involves using Acceptance and Commitment Therapy (ACT) to change the mindset of individuals who internalize stigma or who develop stigma stress as a result of facing public stigma (Fung, Liu, and Wong 2022; Kao et al. 2024). Acceptance and Commitment Therapy promotes mindfulness, psychological flexibility, and a sense of agency and empowerment (Fung, Liu, and Wong 2022). Psychological flexibility challenges a fixed mindset so individuals can perceive challenges in different ways and see and activate different solutions to problems. This helps them cope with experiencing public stigma and makes resisting self-stigma easier. In promoting mindfulness, psychological flexibility, and a sense of agency and empowerment, ACT reduces the disconnection, isolation, marginalization, and powerlessness that many people with mental illness feel due to their illness and how they are treated by society (Fung, Liu, and Wong 2022, 496).

Philosophical Interventions

Philosophical interventions that include reflection, assessment, and introspection can be helpful in combatting self-stigma. Philosophical intervention involves learning how to reason more effectively and how to assess beliefs, values, and reasons for action, as well as learning how to assess one's worth, competence, and the scope of one's agency better. Philosophical development can help a person be more reflective about their beliefs, desires, and values, and decide for themselves which of these should be endorsed and followed. This in turn can help them be more reflective about the role that negative stereotypes play in their lives.

To avoid internalizing negative stereotypes, a person has to be able to assess the validity of the stereotypes and the legitimacy of applying them to themselves, using reason and evidence. The person should be able to examine all the available evidence and reasons, not just viewing a few that seem most salient selectively. The goal is to see how stereotypes are either blatantly false or overgeneralizations that miss the nuance and complexity of lived experience, including the experience of the person themselves. For example, the belief that a person's difficulty with functioning in some area reflects global incompetence should be reevaluated to take into account the complexity of a person's experience and the way problems with functioning in some areas occur alongside the ability to function well in other areas. Assessing the validity of stereotypes requires reevaluating beliefs so that they better reflect reality and are not false or simple overgeneralizations.

Along with assessing the validity of stereotypes, a person should be reflective about which social norms are worth adopting and which should be abandoned. Social norms that are too restrictive and that harm well-being should be abandoned, and social norms that validate a person's experience and increase well-being should be adopted. For example, a person should reassess the value of social norms idealizing independence, self-sufficiency, self-mastery, and personal responsibility, and replace these with social norms of interdependence, mutual needs, and shared responsibility. Reconsidering what social norms should be adopted allows a person to avoid developing adaptive preferences that are oppressive, otherwise harmful, irrational, or nonautonomous. Moreover, if a person can revise the social norms they endorse to better support their well-being, they do not have to be afraid of not living up to social norms that seem by society to be valuable but are actually harmful to the person, helping them to avoid stereotype threat.

Belongingness

Another type of intervention that makes it easier to cope with public stigma and also hinders the internalization of stigma is having a positive view toward one's social group and a sense of belonging (Chan, Mak, and Lam 2018; Elliott and Doane 2015; Seo, Lee, and Lee 2024). Any amount of stigma, even mild stigma, can impair a person's sense of belonging, but people who have a greater sense of belonging and a greater sense of perceived recovery are less likely to have high levels of stigma (Drapalski et al. 2023). Having a positive view toward one's social group involves valuing the group, identifying with the group, and having a sense of belonging and community with the group (Chan, Mak, and Lam 2018). Identifying with one's social group consists of having a person's self-concept defined at least partly by their membership in the group; this is a cognitive process of developing a relational self-concept (Chan, Mak, and Lam 2018). The sense of belongingness and community with the group is an affective experience that contributes to a feeling of being valued by others, to having others in the group help meet the person's social needs, to connecting emotionally with others, to feeling empathy toward others, and to receiving empathy from others (Chan, Mak, and Lam 2018). Having a relational self-concept where a person identifies with a social group that they feel a sense of belonging and community with increases well-being and self-esteem, increases peer support, and protects against internalizing negative views toward oneself such as what occurs with self-stigma (Chan, Mak, and Lam 2018; Elliott and Doane 2015).

Of course, feeling a sense of belongingness within a social group does not always lead to decreased stigma when the social group is itself stigmatizing. As discussed in Chapter 1, when a social group is particularly stigmatizing of mental illness, such as is the case with some marginalized racial groups, feeling a sense of belongingness within that social group can lead a person to feel more stigmatized by their group, which can also lead to increased internalization of stigma (Misra et al. 2021). For example, as noted in Chapter 1, some Asian cultures value in men strength, stoicism, self-sufficiency, and being a provider, which can lead Asian men to perceive mental illness, and receiving treatment for mental illness, as a weakness (Morrow et al. 2020). In another example, also discussed in Chapter 1, in some Latinx cultures, people with mental illness are often seen as dangerous and out of control, and mental illness is viewed as permanent and incurable; these beliefs commonly result in stigmatizing attitudes toward people with mental illness within these communities (Helu-Brown, Barrio, and Hernandez 2023).

At the same time, not everyone within a social group that holds cultural values and beliefs that stigmatize people with mental illness feels stigmatized by their social group or is subject to prejudice or discrimination. Many people within these communities also feel accepted, supported, and nurtured by their communities, especially in comparison to how they feel they are treated within the dominant culture. Having a sense of belongingness within a person's social group can decrease stigma when they have positive views toward their social group and their social group is a source of understanding, empathy, compassion, acceptance, and nurturing.

Willful identification with a social group thus only increases well-being when a person has a positive regard for that social group. When they feel negatively about the social group of which they are a member, they are more likely to conceal that part of their identity and may be more prone to being stigmatized and discriminated against. Concealment and discrimination decrease willful identification with one's social group and diminish a person's sense of belonging and community, which in turn decreases well-being and enables self-stigma to occur more easily (Elliott and Doane 2015). For this reason, it is important that people find favorable aspects of their social group that enable them to have a positive regard for the group and their membership in it.

Support

Having various kinds of support from a person's social group, their family and friends, and the social institutions and systems that they are part of also helps reduce stigma (Adu et al. 2021; Elliott and Doane 2015). Having support and social networks increases a person's well-being and self-esteem, making it easier to cope with public stigma and easier to resist the internalization of stigma. Support can include emotional support, financial support, and practical support such as helping individuals to access treatment or getting their basic needs met. All kinds of support fortify a person's reserves so they have the energy and ability to cope with and resist stigma.

The support of peers is especially important (Firmin et al. 2017, 188–190; Swistak 2024, 543). People with mental illness can gain peer support through participation in peer support groups such as National Alliance on Mental Illness, Depression and Bipolar Support Alliance, Active Minds (for young adults), and the Hearing Voices Network. Peer support lessens stigma by helping individuals to meet and get to know other people who have similar diagnoses. With peers, they can share their experiences in a safe and supportive atmosphere. This kind of sharing and support instills hope in being able to recover and/or live a more fulfilling life.

Self-Disclosure

Self-disclosure of mental illness can be helpful in combatting stigma (Adu et al. 2021; Ege and Lannin 2023; Elliott and Doane 2015; Firmin et al. 2017). Self-disclosure can be positive or negative, depending on how a person feels about the social group they are disclosing their identity in relation to, and depending on what type of reaction they get from the people to whom they disclose (Ege and Lannin 2023; Elliott and Doane 2015). When they feel more positive about their social group, they tend to have a better experience with self-disclosure than when they feel negatively toward it, and when they receive supportive responses from the people to whom they disclose, their disclosure is a more positive experience than when the reactions they receive are unsupportive or stigmatizing. Strategic disclosure in peer support group settings can be particularly effective because these settings are especially supportive (Qin et al. 2024).

Disclosure can thus either decrease stigma through increasing support, well-being, and self-esteem, or increase stigma when others react poorly or when individuals feel unsupported and their well-being and self-esteem are challenged. However, disclosure can be an important strategy for resisting stigma under the right circumstances. People who self-disclosure often perceive themselves as experiencing less stigma and greater social support and self-esteem (Silván-Ferrero et al. 2022, 121).

Challenging Stigma through Advocacy

Participating in advocacy work can be a meaningful way to resist stigma. Theorists identify three main forms of resisting stigma: concealment of one's own diagnosis, deflection of stigma so that it seems to others to not relate to the person, and challenging stigma by showing others how wrong and wrongful it is (Marcussen, Gallagher, and Ritter 2021, 21). Resisting stigma through actions that challenge it increases a person's agency, which consequently improves their well-being (Marcussen, Gallagher, and Ritter 2021, 21). Actions that challenge stigma can be thought of as advocacy work.

Advocacy work can be performed at an individual level or through collective action. One form of advocacy is educating others about mental illness to increase mental health literacy, both in an interpersonal context and through social means such as campaigns that try to replace myths with facts (Firmin et al. 2017, 190). Education also involves challenging stereotypes with counter-examples and arguments about why stereotypes are false or falsely applied. Another form of advocacy is confronting others about their biases and prejudice to encourage them to reflect on and change these (Firmin et al. 2017, 190). Advocating for others can also be an important component of advocacy work (Firmin et al. 2017, 191). Collective action to dismantle systems of oppression that perpetuate stigma, such as through boycotts, lockdowns, or demonstrations, is also important. Individuals can perform some of these actions on their own, but such actions are often more effective when they are undertaken by collective organizations through collective actions (Garcia-Lorenzo, Sell-Trujillo, and Donnelly 2022; Howarth 2006, 448).

Individual Interventions for Others in Addressing Public Stigma

Interventions to address public stigma include some of the same psychological interventions that help individuals subject to stigma cope with public stigma and resist self-stigma, such as mindfulness and ACT.

Philosophical interventions involving reflection on and reassessment of negative stereotypes, implicit bias, and oppressive social scripts is critical, as in narrative engagement, reconceptualizing mental illness and recovery, and reassessing social norms. Education and interpersonal contact can be helpful in the right circumstances. Building communication and interpersonal advocacy skills is a crucial way to address stigma in institutional settings where people with mental illness are clients or colleagues. In addition, collaboration with people who have mental illness is vital in developing policies and practices that resist stigma and are effective at benefitting people with mental illness. Developing a better understanding of mental illness, mental illness experience, and recovery can improve empathy, enabling more supportive and less stigmatizing attitudes and behaviors toward people with mental illness.

Psychological Interventions

Psychological interventions can help people in the general public decrease the stigma they project onto others. When people who have a propensity or possibility of stigmatizing others engage in mindfulness, they decrease the rumination and negative feelings that feed stigma, making it easier for them to have the psychological flexibility that helps them avoid projecting stigma. Acceptance and Commitment Therapy is also beneficial in reducing stigma in the general public (Fung, Liu, and Wong 2022; Kenny and Bizumic 2016). When individuals with a propensity or possibility of stigmatizing people with mental illness engage in ACT, they develop the psychological flexibility to understand the lives of people with mental illness better, which increases empathy and decreases stigma. In fact, ACT has been found to be more effective at reducing stigma than education about mental illness (Kenny and Bizumic 2016). In addition, having a growth mindset (rather than a fixed mindset) that emphasizes that a person can grow through learning and being willing to learn more about mental illness leads to less stigma (Babij et al. 2025).

Philosophical Interventions

Friends, family, people who work with those who have mental illness as clients, and people in the general public can benefit from philosophical interventions as well. One important intervention involves deep introspection, either within the person themselves or intersubjectively with others, to recognize the roles stereotypes play in their thoughts, emotions, and behavior so that they do not impose stereotypes on vulnerable individuals. In addition, addressing public stigma requires that individuals become aware of implicit biases they hold and actively disavow them as well as work hard on resisting and changing them. When interacting with people who have mental illness, they need to continuously repudiate negative stereotypes that circulate within the culture regardless of their relevance in a given interaction. They should also become aware of cognitive biases they might hold that support negative stereotypes so they can try to avoid employing these biases.

While they need to resist explicit stereotypes and implicit biases, individuals trying to avoid imposing public stigma also need to recognize and reevaluate oppressive social scripts about people with mental illness, including the motivations and expectations that are attributed to them. This involves reevaluating their beliefs about the causes of mental illness so they have a deeper and more nuanced understanding of the complicated and intertwined roles that biological, neurochemical, psychological, and social factors play. This also involves recognizing that all individuals, including people with mental illness, have nuance and complexity in their experience, and that their experiences and actions cannot and should not always be reduced to being about their mental illness. Instead, individuals need to develop a deeper understanding of the myriad causes of mental states and behavior and recognize with epistemic humility that they often do not know – and do not always need to know – what the causes of a person's motivations, thoughts, feelings, and behavior are.

Other philosophical interventions that can decrease stigma include narrative engagement and reconceptualizing mental illness and recovery (Heney 2022). Narrative engagement involves paying attention to the person's lived experience with mental illness by engaging with their narrative, including both the narrative that others ascribe to the person and the narrative the person gives of their own experience. In fact, epistemic justice demands that others prioritize the narrative the person gives themselves because they are the expert in their first-person knowledge of their experience, and others should have the epistemic humility to recognize the limits of their own knowledge and be open to what they can learn from the self-narrative of the person with mental illness. Narrative engagement can help someone see the nuance and complexity of the experience of a person with mental illness so they can see the person as a unique individual with unique experiences that cannot be reduced to their illness. Narrative engagement can also help someone see what social relationships matter to the person with mental illness, how they identify themselves, and what social groups they feel they belong to. This helps people to avoid de-individuating and mis-identifying individuals with mental illness.

In addition, recovery should be reconceptualized as involving the development or restoration of capacities that mental illness has impaired, which increases a person's agency and autonomy and rehumanizes them in the eyes of others. When others dehumanize the person with mental illness, they believe that mental illness makes a person less human. Reconceptualizing mental illness as causing temporary impairments that are reparable and as being a temporary state that a person can develop and grow from can help others reconsider what it is to have mental illness. Reconceptualizing recovery as the development or restoration of capacities can help others see recovery as a journey out of temporary and reparable impairments that can help a person flourish as a human being and thus rehumanize them. Narrative engagement and reconceptualizing mental illness and recovery are important methods of decreasing mental illness stigma in areas of healthcare, work environments, family and social settings, and the general public.

Finally, reassessing social norms about what an ideal individual in a twenty-first-century capitalist, democratic society should be like is also critical. Norms that idealize independence, self-sufficiency, individual action, self-control, self-mastery, and personal responsibility should be reassessed and reconsidered in light of norms about the value of interdependence, mutual needs, collective action, and shared responsibility. Social norms that harm vulnerable individuals should be reevaluated and replaced with norms that are more relational and supportive and that enable the growth and flourishing of all.

Education

Education about a stigmatized trait and identity such as mental illness can sometimes be another valuable intervention to defuse stigma, especially when it focuses on increasing mental health literacy and normalizing discourse around mental health and illness (Firmin et al. 2017; Hejizadeh 2024, 9; Na et al. 2022). Some studies of anti-stigma educational programs have shown that increasing mental health literacy through education can counter stigma (DuPont-Reyes et al. 2023; Waqas et al. 2020). Diana Heney defines education as the production of positive new attitudes (Heney 2022, 885). Having more knowledge about mental illness can enhance empathy, decrease stigmatizing attitudes, and increase conscious changes in action to combat stigma (Na et al. 2022, 136).

However, education does not always effectively lead to less stigma. One study showed that in the short term, both ACT and education made people less compassionate and supportive compared to their states before the intervention. The authors of the study suggested that perhaps the emphasis given with these interventions that people with mental illness are no different from others reduced other people's inclination to help and support them (Kenny and Bizumic 2016, 183). Moreover, another study showed that education about mental illness can backfire when what is presented as myths (false beliefs about mental illness) is heard and internalized by others as facts (true beliefs about mental illness) (Dobson and Rose 2022). In addition, campaigns that object to ads that tie themes of mental illness to a product (such as wild stereotypes of madness and lunacy to depict excessive desire for the advertised product) can backfire in further cementing the association between mental illness and extreme behavior and desire and leading to increased stereotyping of what it is to have mental illness (Holland 2012).

Patrick Corrigan argues that a proposed "educational fix" often does not work (Corrigan 2018). When education includes detailed instructions, challenging implementation, and/or occurs over a long period of time, it can overload people with information and demand too much of them, making it hard for them to process what they are learning and what they are supposed to do. At the same time, when education is discrete (one-time), brief, and to the point, it can be too limited to be useful, also making it hard for people to process and utilize effectively. Because information can be difficult to process and implement either way, it can create a "boomerang effect" of psychological resistance, which is the opposite of the desired effect of creating a growth mindset and giving people tools they can use effectively. While some studies suggest that education can be a valuable way to reduce stigma, it is not always effective or even helpful; its effectiveness probably depends on the context in which it is employed.

Interpersonal Contact

Interpersonal contact, which occurs when a person with mental illness is willing to self-disclose, and which often involves a person with mental illness sharing their story, is another important intervention to reduce stigma (Adu et al. 2022; Alexander and Link 2003; Heney 2022, 886; Na et al. 2022, 136). The *Lancet* Commission on Ending Stigma and Discrimination in Mental Health found that interpersonal contact with people who have mental illness was the most effective way to defuse stigma of all methods studied (Thornicroft et al. 2022). Personal contact with people who have schizophrenia, for example, leads to more positive attitudes toward them, less fear and anxiety around them, less desire to avoid them, and less stigma in general (West, Hewstone, and Lolliot 2014). Contact-based stigma-reduction programs have been shown to be effective in reducing stigma in many populations, especially Asian and Latinx populations (Wong et al. 2018). One study of virtual storytelling of lived experience through video reduced stigma in all populations studied and especially in Black Americans (Conner et al. 2024). When individuals share their story, it allows others to have a better understanding of the lived experience of people with mental illness.

However, one study showed that interpersonal contact with someone who had mental illness did not reduce implicit bias against mental illness (Robb and Stone 2016). Contact with a person who has mental illness can actually increase stigma when the person with mental illness is not prepared to address the stereotypes and prejudice aimed at them (Dobson and Rose 2022, 27). Interpersonal contact is thus not a panacea, either, but it can be effective when paired with other interventions that increase understanding and empathy of mental illness experience and recovery.

Skills Building

Building skills to address stigma is important, especially in institutional settings like healthcare, workplaces, and housing and disability service centers (Kachur and Harter 2024, 357; McAllister et al. 2023). One type of skill that can be developed is communication skills that help people avoid committing as well as overtly address microaggressions and other displays of stigma. Another type of skill is interpersonal skills of building understanding, empathy, trust, and rapport with people with mental illness, whether they be clients or colleagues, as ways of interacting with them in more supportive and less stigmatizing ways. Allyship and skills of advocacy are also important to align oneself with those who are marginalized and to work for changes that aid the marginalized.

Collaboration

Collaboration with people who have lived experience with mental illness in creating policies and practices at governmental and institutional levels is crucial for reducing the misunderstanding and paternalistic approaches that can be involved when decision-making excludes these individuals.

Studies of reducing stigma in the medical and mental healthcare systems show that collaboration with healthcare service users is essential to increase understanding of and empathy toward people with mental illness, and to promote agency- and autonomy-promoting interventions (Bobbili et al. 2022; Hejizadeh et al. 2024, 8). Collaboration with service users is also essential within systems that serve people with mental illness, such as housing and disability service centers, in order to create policies and practices that respect and truly benefit people with mental illness rather than promoting paternalistic, harmful, and stigmatizing interactions based on psycho-pathophobia. Giving people with mental illness a say over what happens to them within various governmental and institutional settings is an important way to avoid the epistemic and other injustices associated with stigma.

Structural Interventions for Social Institutions and Systems to Address Stigma

Structural responses to address stigma are the responsibility of society to carry out through its social institutions and social systems that contribute to structuring the lives of people with mental illness and the general public. Some structural interventions involve helping individuals with mental illness to have access to various resources and opportunities, such as finding and keeping jobs and acquiring and maintaining living spaces. Other structural interventions include accommodating the needs, interests, ideas, and contributions of people with mental illness so that these are valued within society, and affirming the dignity and worth of people with mental illness.

Social institutions and systems need to help individuals with mental illness get their basic needs met so they can participate as equals in society. Some of these basic needs include helping people to find and maintain employment, assisting them with acquiring safe and stable living spaces, ensuring they have access to food, and enabling access to adequate healthcare. It is paramount that people with mental illness have access to the same resources and opportunities that others have access to, not only the basic needs necessary for participation in society but also other resources and opportunities that are available to others who do not have mental illness, including resources available to the middle class and opportunities to carry out different social and vocational roles. Not only does this enable people to participate in society in equal ways compared to others but it also allows them to have more options. As noted in Chapter 6, having multiple good options is crucial for developing and exercising autonomy.

Various social structures and institutions need to change how they are organized in order to better accommodate the needs, interests, ideas, and contributions of people with mental illness. For example, healthcare systems, including hospitals and behavioral health clinics, need to implement methods for obtaining feedback from the people they serve about how well they are doing and what changes they should make. Housing, job placement, and disability services need to solicit feedback from their clients about what their clients' needs are and how they can better serve their clients. In addition, institutions should make an effort to recognize and honor the contributions of people with mental illness, in part by using their contributions to better society.

Social structures and institutions should be organized in a way that focuses on the capacities rather than deficits of people with mental illness by encouraging and enhancing areas of strength. While some institutions are designed to address the various mental health and functional challenges of people with mental illness, such as hospitals and behavioral health centers that provide treatment and recovery services, they can do so in a way that draws on people's capacities and strengths as a way of dealing with and overcoming incapacities and limitations. This is similar to a movement within special education of focusing individual education plans on capacities and strengths rather than deficits (Elder, Rood, and Damiani 2018).

With a relational conception of the self, social structures and institutions should welcome and draw upon the interdependence, mutual needs, collective action, and shared responsibility of vulnerable individuals as sources of strength. They should value these norms rather than the norms of independence, self-sufficiency, individual ability to get their own needs met, individual action, and personal responsibility. This would enable vulnerable individuals to thrive within a system that supports them rather than relegate them to uselessness in a system that precludes them.

Social institutions and systems need to affirm the dignity and worth of people with mental illness. Recognizing people's dignity and validating their worth is not only the job of particular individuals who interact with people with mental illness; it is also a structural issue that needs to be carried out within social institutions and systems that structure the lives of people with mental illness. Treating people with mental illness as epistemic and moral agents who know what is good for themselves and who can carry out responsibilities and interact with others in meaningful ways is an important way of honoring their dignity and treating them as having intrinsic value.

Let us examine more closely two areas where structural interventions are paramount for people with mental illness: employment and housing.

Employment

Many people with mental illness have a difficult time participating in employment for many reasons. One study identified lack of motivation to follow through on commitments to employment services, lack of encouragement and support from others, criminal history, and substance use as factors (Sveinsdottir and Bond 2020), while another study identified symptoms, medication side effects, nonadherence to treatment (including medication), communication impairments, lack of social skills, and difficulties with following instructions and carrying out role responsibilities (Nirmala et al. 2020). However, as explained in Chapter 1, another factor that prohibits employment for many is discrimination. Many people with mental illness could do various kinds of work if they were given the opportunity, but they are often denied this opportunity due to negative stereotypes about what people are capable of.

People with mental illness often have positive skills and abilities that can be conducive to productive employment. These include developing a strong work ethic (from working hard on recovery), coping skills, having different and unique perspectives, and being able to consider alternative approaches; many of these abilities arise from living with and coping with mental illness and active engagement in recovery. When there is a good fit between their abilities and the demands of a job, people with mental illness can thrive in the workplace and experience positive self-worth and other self-regarding traits as well as improved mental health and well-being (Hennekam, Follmer, and Beatty 2021). When there is not a good fit between a person's abilities and the demands of a job, however, a person can experience lowered self-esteem and decreases in positive self-regarding traits like self-confidence, self-trust, and self-respect. Mental illness symptoms can interfere with a person's ability to perform a job well when a person has difficulties with concentration, emotion regulation, communication and social interaction, or coping with stress (Hennekam, Follmer, and Beatty 2021, 3256). It is paramount for employers and employees together to identify positive skills and abilities that the person with mental illness has and to match these with the requirements of a job so there is a good fit. When an agency helps people who have mental illness with job placement, the focus should be on their capabilities, not their deficits, so that the goal is to match individuals with jobs where the individual is capable of doing the work involved and can bring identifiable assets to the position.

Many people with mental illness who work do so in the context of supported employment. Supported employment consists of placing individuals with disabilities in work programs that are more flexible and have easier role responsibilities with which people with disabilities can be successful. Evidence indicates positive results for people with mental illness who engage in supported education and supported employment (Schindler and Sauerwald 2013). Supported employment often occurs in social enterprises, which are organizations that have as their primary aim to build social capital and meet the needs of people who are underserved by society; these tend to consist of non-profits or worker cooperatives (Wilton and Evans 2018, 562). Social enterprises are usually better at accommodating the needs of people with mental illness, such as with schedule flexibility, work-task flexibility, flexibility in what social interaction is required, job security (even in the face of absenteeism, lateness, or time off for dealing with illness), and other social support such as housing and disability benefits (Wilton and Evans 2018, 568–571). In addition to benefitting people with mental illness, supported work programs are economically advantageous for communities, offsetting the cost of treatment (Booth et al. 2014).

Other kinds of employment support include placing individuals in competitive work positions and supporting them through accommodations, encouragement, and counseling to be able to persist in their employment. Individual placement and support services are successful at helping people to maintain their employment over time (Johnson-Kwochka et al. 2017). Success in a competitive work environment tends to require people with mental illness to exhibit self-confidence and self-trust in their abilities and commitments (Gladman and Waghorn 2016, 841). In a virtuous cycle, being engaged in meaningful work also fosters these. People with mental illness who work at jobs that are meaningful to them have a more positive self-concept and more positive self-regarding traits, including self-esteem and self-efficacy, as well as greater empowerment and well-being (Waghorn and Lloyd 2005, 13).

One issue that people with mental illness often have to face in the workforce is whether to disclose their illness to their supervisor and/or colleagues. Many people with mental illness are reluctant to self-disclose in the workplace because they fear that they will experience stigma and/or lose their job (Donnelly 2017). Indeed, sometimes upon disclosure people with mental illness receive increased support, but sometimes this strains their work relationships (Jones 2011). Disclosure is more commonly made to supervisors than to coworkers (Jones 2011). Some employers welcome employees and prospective job seekers to disclose their mental illness, seeing their honesty as an exhibition of courage and integrity; this is especially the case when employers already have close contacts with people who have mental illness and are familiar with their experience (Østerud 2023, 100–102).

Employers can play an important role in reducing mental illness stigma and aiding employees who have mental illness to help buffer whatever stigma they do face. Employers can reduce mental illness stigma by providing educational seminars about psychiatric and other disabilities and providing informational brochures (Janda 2008, 438–439). They can help workers who have mental illness by making counseling services accessible, providing adequate health insurance so that healthcare is more accessible, providing flexible work hours, and responding to environmental stress with changes in lighting and sound (Janda 2008, 440–445). Empowering employees in the workplace and providing beneficial work arrangements is an important way of supporting employees with mental illness and diminishing workplace stigma (Hejizadeh et al. 2024, 9). By supporting employees who have mental illness, employers can help them cope with workplace stress and deal with and avoid stigma they might encounter in the workforce.

Housing

Many people with mental illness also need support in finding safe and stable housing. This involves both moving homeless people into permanent housing and helping people in precarious housing situations find decent apartments (clean, properly maintained, in safe areas, and stable over time) where landlords do not discriminate against them. Helping to match tenants with apartments can require focusing on the capabilities of the person with mental illness rather than their deficits to show landlords what good tenants people with mental illness can be. Helping individuals to maintain their living conditions over time requires providing support to people with mental illness to help them be better tenants, such as by making therapy and medication more accessible and affordable, and having support people such as counselors and social workers available 24/7 to help with mental health issues and crises.

The traditional way of sheltering people who are homeless, many of whom also have mental illness, is through traditional housing programs that take a "staircase" approach (Padgett, Henwood, and Tsemberis 2016, 7). This starts with putting people into emergency shelters and then moving them into transitional housing (halfway houses) where they are supervised by staff, with the possibility of later moving into more permanent housing such as Section Eight housing, which gives people financial help to afford low-income housing. In this approach, both transitional and permanent housing are conditional on people participating in mental health and/or substance abuse treatment and on people having to prove they are ready to move on to the next step. This approach can be helpful for homeless individuals to gradually work their way into permanent housing.

A newer alternative method is the Housing First approach, which provides housing with no conditions attached (National Alliance to End Homelessness 2016). Housing First usually provides mental health and substance abuse treatment near where individuals live, and they encourage individuals to participate in it, but they do not require this participation as a condition of housing. The Housing First approach gives people choices about where they want to live and provides scatter-site housing that allows individuals to live within the broader community so they live with a broader range of people with different backgrounds, not just with formerly homeless people like themselves (Padgett, Henwood, and Tsemberis 2016, 3).

Many studies have demonstrated the success of Housing First programs in getting people housed stably over time; housing stability rates for Housing First programs range from 77 percent to 88 percent (Padgett, Henwood, and Tsemberis 2016, 50 and 56–57). Moreover, people in Housing First programs tend to take advantage of mental health and substance abuse treatment when these are provided as support in conjunction with their housing (Tiderington et al. 2020; Whisler et al. 2021). They also have better medication adherence and often discontinue or avoid illegal drugs; people in Housing First placements do not have increased mental illness symptoms, suggesting that they willingly take prescribed medication at the same rate as people in Treatment First programs, and they also do not use drugs at a higher rate (Padgett, Henwood, and Tsemberis 2016, 54–56). As a result, they have reduced need for inpatient psychiatric care (Kerman et al. 2018), fewer hospitalization stays (Dunt et al. 2022), and lower rates of incarceration (Leclair et al. 2019; Luong et al. 2021). People in Housing First programs use these services at lower rates than those who are not in such programs because their lives are more stable (Culhane, Metraux, and Hadley 2002; Dunt et al. 2022; Kerman et al. 2018; Loubière et al. 2022). One consequence of these lower rates of service is that Housing First programs save public money over time, since paying to house someone in their own apartment is less expensive than paying to have them use emergency rooms, hospitals, and jails. Another consequence is better quality of life for those who participate in Housing First programs, including better safety, enhanced family and social relationships, and the ability to pursue work as well as engage in leisure activities (Patterson et al. 2013; Henwood et al. 2014).

Social and Cultural Interventions to Address Public Stigma

Other types of interventions that are necessary for addressing public stigma are social inclusion and interaction, and cultural reassessment of social norms. Social interventions involve promoting social inclusion and enabling and encouraging social interactions for people with mental illness, both with others in their social group and with those who are not in their social group. Addressing cultural impacts on stigma requires philosophical interventions at the cultural level involving reassessment of values and social norms, and changing cultural understandings of mental illness, mental illness experience, and recovery.

Social Interventions: Social Inclusion and the Availability of Social Interaction

To eradicate stigma, it is crucial that people with mental illness have a sense of belonging in their communities, both the social groups of which they are members and society at large. Having a sense of belonging involves being accepted as a legitimate, credible, and authoritative member of society who is viewed as an equal participant in moral, epistemic, and social practices of the community. It involves being seen as someone who can receive and respond to moral claims by others and make moral claims of their own on others that will be respected and responded to appropriately. It also involves being seen as someone who can share ideas and knowledge that are taken seriously by others and responded to in uptake, and who can respond to others' ideas and knowledge as credible interlocutors. Finally, it involves being seen as someone who can interact with others appropriately in their speech, emotional attunement and responsiveness, and behavior.

Being accepted and viewed in these ways is critical for someone to have a sense of belonging in their community.

Social interaction of various kinds is thus important to develop this sense of belonging. Individuals, social institutions, and systems alike all have the responsibility of enabling and encouraging social interaction of people with mental illness. Individuals increase social interaction through warm and supportive individual behaviors around people with mental illness. Social institutions and systems foster social interaction through structural, systemic changes that are designed to smooth the way for people with mental illness to interact more often and more easily both with others like themselves and with people who do not have mental illness.

One kind of social interaction people with mental illness need is interaction with others who are like them, in what we can think of as *homogenous social interaction*. When they interact with others in their social groups, including others with mental illness, but also others who are of the same race, gender, socioeconomic status, and ability status, they can get important needs met through community engagement like acceptance, validation, empathy, and various kinds of support. When they connect with others who are members of the same group, they derive strength, support, and meaning. They can empathize with others who are experiencing something similar to themselves, and they can receive empathy and compassion from others, creating bonds of connection and nurturance. In the context of mental illness, they can find meaning in their experiences of mental illness that allows them to grow from the experience, develop new abilities and coping mechanisms, and form relationships with others. There are many places in their lives where such engagement can occur. Support groups, living situations, healthcare settings, and sometimes jobs can be important places where individuals with mental illness interact with others who have mental illness, which can provide them with a valuable sense of belonging, community, and support.

It is also important, however, that people who have mental illness interact with people who are unlike them, who do not have mental illness, so they can benefit from different kinds of support as well as resources and opportunities they would not otherwise have access to. This occurs through *heterogenous social interaction*. When people interact only with others who are like them, they are often limited to what resources, opportunities, and support they have access to, namely, those that they might already have access to given their social location. When people interact with others who are unlike them, on the other hand, this can open doors to other resources, opportunities, and support that they do not normally have access to given their social location. For this reason, individuals, social institutions, and systems must enable and encourage social interaction between people who have mental illness and those who do not. This allows people with mental illness to have a heterogeneous range of contacts and interaction, not just the homogeneous range where they only interact with people like themselves.

Individuals, social institutions, and systems all need to affirm the belongingness of people with mental illness explicitly so there is no uncertainty about whether they belong. This includes belonging both to one's own social groups, including the group of people who have mental illness, as well as to the broader community as a whole. Feeling connected to one's social group, even when it is a social group that experiences discrimination, mediates internalized stigma and raises self-esteem (Silván-Ferrero et al. 2022), which improves self-regarding traits such as self-trust, self-confidence, self-efficacy, and self-respect. Feeling connected to the broader community also accomplishes these ends.

Affirming the belongingness of people with mental illness is an important way to address stereotype threat. Claude Steele recommends affirming the belongingness of people subject to stereotype threat within the relevant domain where stereotype threat applies (Steele 1997, 625). In the case of African American children, who were the subjects of his study, this means affirming their belongingness in the academic domain so that they can feel a sense of belongingness rather than judgment and ultimately rejection. In the case of people with mental illness, this means affirming their belongingness in general society, as one of "us" rather than as one of "them" so as to minimize the sense that they are fundamentally different from others. This requires other people in society to focus on the similarities they share with people with mental illness, to see the humanity in people with mental illness, and to seek out greater understanding of the experiences and perspectives of people with mental illness. When other people can build "empathetic bridges" with the person with mental illness (Davidson 2003, 121–125), they can humanize them and see them as simply another person like themselves.

This also requires other people to minimize the significance of differences they see between themselves and people with mental illness. One difference, for example, is seeing people with mental illness and other disabilities as dependent on other people and on government and nonprofit systems that give them aid to be able to get their basic needs met and to participate in daily living activities. This is in contrast to social norms that idealize independence, self-control, self-mastery, self-reliance, self-sufficiency, individual action, and personal responsibility. This seeming contrast can make some people very judgmental of those who are dependent on others. Affirming the belongingness of people with mental illness in the domain of general society requires minimizing the moral significance of dependence so that it is not seen as a salient feature that divides people who are successful and who belong (one of "us") versus people who can't make it on their own and don't belong (one of "them"). It requires understanding dependence as interdependence and seeing that even people who seem to not need the help of others to get their basic needs met and participate in daily activities are nonetheless interdependent on others for various kinds of resources and support. It involves a transformation of how we understand the moral significance of dependence and interdependence.

Affirming the belongingness of people with mental illness also involves supporting their belongingness in the group of people who are competent and who can cope with their problems. This can be done by focusing on the capabilities a person with mental illness has rather than focusing on their disabilities and mental and behavioral impairments. Emphasizing, drawing upon, and further fostering the capabilities a person has supports their feelings of competence and their confidence in their ability to cope, which in turn supports their sense of belongingness in the group of people who are competent and can cope with problems. In *feeling* more competent, and having more confidence in one's abilities, a person actually *develops* more competence because they employ their confidence and feeling of capability toward the capacities involved with competence. Confidence and a feeling of capability allow a person to think through options more rationally, to consider options they might not have considered if they lacked confidence in themselves, to be more motivated in trying to accomplish their goals, and to work harder toward that end. Emphasizing, drawing upon, and further fostering a person's capabilities allows them to become more capable, and thus more competent, which both defeats the salience of the stereotype of incompetence (Benson 2015, 131) and increases a greater sense of belonging in the domain of competence and coping.

The belongingness of people with mental illness in various communities should be firm and explicit. People with mental illness need to be treated as epistemic and moral agents who can participate equally in epistemic, moral, and social practices of the community and whose participation is recognized as credible, legitimate, and authoritative. They need to be explicitly recognized and treated as members of a community who are equal in all meaningful ways as other members. There should be no doubt in the mind of a person with mental illness about whether they belong to the community based on how they are treated. They should be accepted and validated in a way that lets them know with certainty of their belongingness in their communities.

Being an epistemic and moral agent and having normative authority require being recognized by others as an agent and as having normative authority. This recognition occurs in the context of a community that can validate a person's worth and enable the person to participate in the moral, epistemic, and social practices of the community. Through this participation, they are treated as equal compared to others. In addition to enhancing their agency and normative authority, this also helps the person with mental illness to develop a more positive self-concept and increases self-regarding traits like self-trust, self-confidence, self-efficacy, and self-respect, all of which help buffer stigma. Moreover, recognizing and valuing the agency and normative authority of an individual helps to resist imposing status loss and dehumanization on them so they are valued as full human beings with intrinsic worth. Individuals within a community need to recognize the agency and normative authority of people with mental illness, and social structures need to be organized to facilitate this recognition and validation.

In addition to recognizing and valuing the agency and normative authority of people with mental illness, individuals and social institutions alike need to recognize and value the relational self. They need to see and treat a person with mental illness as a unique individual with nuanced and complex experiences that are not reducible to mental illness. This individuates the person and prevents their identity, thoughts, feelings, and behavior from being reduced to their social trait of having mental illness. Community members and institutions also need to recognize and respect the various social groups that individuals with mental illness belong to correctly. This helps identify them with respect to their relational selves more accurately and thus affirms their relational identity through social validation. This is yet another way that the belongingness of people with mental illness is explicitly upheld.

Philosophical Interventions

Philosophical interventions of cultural introspection, reflection, and assessment are important for both individuals in society as well as social institutions and systems. Society needs to engage in cultural introspection about its values and social norms so that it can reassess what is valued in individuals. This includes reassessing the negative worth placed on dependence, neediness, and incompetence so that these can be seen in a more positive light, such as a place of vulnerability that obligates people to give care and concern, and an occasion for people to carry out responsibilities and exercise virtues in helping to care for those in need. If cultural values in these areas changed, this would allow people with mental illness to not be afraid of embodying states of dependence, neediness, and difficulty with functioning. Eliminating this fear would eradicate stereotype threat and fear of being judged, as individuals with mental illness would be less afraid of and anxious about embodying these states.

Changing cultural values requires embracing the relational self. Instead of valuing a self that is economically rational, autonomous, sane, self-controlled, self-regulated, self-directed, responsible for their own actions, able to get their own needs met, independent, self-sufficient, safe, and stable, we need to value a self that is social, relational, interdependent with others, and in relationships of mutual need and responsiveness. We need to accept that everyone has needs that they cannot meet on their own and that necessitate the help of others to get met. Everyone likewise has obligations to help others in need when and where they can. This reciprocity of need and obligation makes us interdependent with others where we are sometimes and in some ways needy and sometimes and in some ways capable.

Part of valuing the relational self is recognizing the importance of social recognition of others as equal members of a moral, epistemic, and/or social community and equal participants in the moral, epistemic, and/or social practices of that community. Affirming people's belonging in the community in explicit and ongoing ways is crucial for them to have the security of knowing they belong. Treating people always as ends in themselves, always as having intrinsic value and moral worth, always as having dignity that must be respected is essential. So is actively engaging with them as equal participants in the moral, epistemic, and/or social practices of the community.

Safety, stability, and being a contributing member of society are also relational. Creating a safe community and a stable future involves shared activity of all community members. Establishing stable dispositions and identity, and being self-regulating, self-controlled, reliable, and trustworthy is a process that occurs in relation to others through one's moral and epistemic engagement with others. We need to recognize that there are other ways that people can contribute to society, help others, and give back than the traditional forms we expect such as through employed work. Caretaking, creating art, sharing self-narratives, engaging others in social and political practices, and helping others to get their needs met are alternative ways of contributing to society.

The self is relational, too, in how it develops and exercises various capacities, including capacities for agency and autonomy. Others with whom a person with mental illness interacts in various domains of their life help the person to develop their various capacities. Other people provide the recognition that a person needs as an equal, credible, legitimate, and authoritative member of society to be able to practice their capacities and develop the skills needed for agency and autonomy. Other people are also necessary for helping a person to be able to see themselves through the eyes of others so they can learn to recognize themselves as equal members of the community and to treat themselves with compassion and empathy.

As with all people, people with mental illness should be seen as having multiple social identities, inhabiting various epistemic positions, and having a plurality of capacities as autonomous moral agents. Replacing the essentialist views that are sometimes directed at people with mental illness with more fluid and plural conceptions is critical. One study that deemphasized a biomedical model of mental illness and thus avoided the essentialist views sometimes associated with a biomedical explanation was particularly effective at decreasing stigma (Dillinger 2021). Which social identity, epistemic position, or moral capacity a person expresses at a given time can vary depending on the context: How they are socially positioned and what the context calls for. But they always have the capacity to be fluid in their expressions of social identity, epistemic position, and moral agency, and their fluidity of expression needs to be recognized and valued by others.

As a society, we need to value the position of being vulnerable as an aspect of being fully human and as an occasion for moral response. Related to this, we need to value practices of care that aid people in their vulnerability, as well as teach people how to care for others better. People with mental illness not only need care in their own position of vulnerability but they also need to be seen as having the capacity to give care to others.

Connected to this, we should value sometimes not being in control and not being self-regulated, and appreciate alternative ways that people think, feel, express themselves, behave, and interact with others. Differences are not always or necessarily disadvantages; they can be assets, too. There can be beauty in the chaos of a person's experience, and gifts like creativity and novelty can arise from the chaos. Differences in thought, emotion, and behavior can open the door to alternative ways of understanding the world and divergent approaches to dealing with issues and problems.

In addition to reassessing cultural values, society needs to change its understanding of mental illness and recovery so mental illness is not viewed as a permanent condition that gets progressively worse over time or as incurable, defective-making, or inferior-making. Both mental illness and difficulty with functioning need to be reconceptualized so they are not seen as permanent and inherently defective but instead seen as temporary states that ebb and flow and that both advantage and disadvantage people in different ways. Mental illness should be regarded as a set of abilities, limitations, and needs that exist on a continuum of different abilities and needs, where all differences are viewed as part of being human, and where differences can change over time. One mental health literacy approach that sought to promote empathy by showing how symptoms exist on a continuum of mental health was particularly effective at reducing stigma (Dillinger 2021).

This is in contrast to viewing mental illness as a set of fixed abilities, limitations, and needs that mark a person as fundamentally different and consequently lesser. Disabilities, including the psychiatric disabilities that mental illness creates, are often viewed as deviations from what is regarded as typical functioning of an organism; psychiatric disabilities are viewed as deviations from a range of what is considered "normal" mental and behavioral functioning. In this "bad difference" view of disability, disabilities are seen as falling outside whatever is considered to be the range of "normal" functioning, and as inherently bad, marking a person as different and lesser (Barnes 2016, 14). Revising our understanding of disabilities, including psychiatric disabilities, as on a continuum of functioning will help us avoid the stigmatizing judgment of people who have disabilities as abnormal, bad, and inferior.

Rather than viewing disability as an outlier of a normal range, we should understand it as on a continuum of "normal" functioning where someone can function well in some areas and more poorly in other areas, where these different areas can overlap, influence each other, and vary over time, and where *all* differences in functioning are seen as on a continuum of normalcy. In this way, mental illness can be seen as a continuum of what it is to be a "normal" human being rather than as marking someone as an outlier, as different and lesser. This treats individuals with mental illness as fully human rather than as being subhuman.

Treating people as full human beings and not dehumanizing them requires accepting that they always have the capacity to be fully autonomous moral agents and fully credible epistemic agents even when these are limited in certain circumstances. Mental illness symptoms and poor social treatment of people with mental illness can both create constraints that diminish a person's autonomy, credibility, and agency, making them have less capacity in practice. It is important to recognize the real limitations that people experience in order to work on addressing them so as to increase their capacity. What is essential is recognizing that they always have the inherent capacity to be full moral and epistemic agents and full human beings despite temporary, circumstantial limitations that they have to deal with. It is when people view those with mental illness as having inherent deficits in agency, autonomy, and credibility that they treat people as fundamentally lesser and dehumanize them.

Reconceptualizing mental illness as temporary states that ebb and flow allows for a more relaxed attitude toward people with mental illness when they are incapacitated and a more encouraging attitude toward them when they are more capable. Rather than seeing people with mental illness as having fixed, inherent limitations that invite people to dismiss them as inherently incapable (incompetent), we should regard those limitations as temporary states where their capabilities and competence can vary. When a person's illness flares up, they may need more rest and recusal from responsibilities, but when they are feeling better, they should be enabled and encouraged to exercise their various capacities. When they are more capable, they can be given more responsibilities and have higher expectations put on them. We need to recognize that capabilities and limitations can ebb and flow, and what we expect of people and what responsibilities a person can handle can change over time. (Assuming that people's capabilities and limitations are variable and can change over time is also important for people who do not have mental illness. It is crucial for *everyone* that we adopt a stance of flexibility and acceptance of differences and changes in capabilities and limitations.)

All of these changes of cultural values, including affirming the many ways that the relational self is significant and reconceptualizing mental illness, require philosophical discernment at a social and cultural level, where people collectively work together to envision and enact a different set of values that is more compassionate and more supportive of vulnerable people such as people with mental illness. This cultural transformation of values is paramount to adequately address mental illness stigma. Only when we change who and what we value will we be able to change the way we treat people: with compassion, empathy, and understanding rather than with stigma.

Conclusion

Addressing stigma involves many different interventions that enable individuals to cope with stigma they experience, to challenge public stigma that is projected onto people with mental illness, and to resist internalizing stigma in the ways discussed in this book. Some of these interventions are discrete actions that individuals can undertake. Others are structural changes that social institutions and systems need to make so they are organized in ways that are destigmatizing and supportive of people with mental illness. In addition, some interventions are social and cultural efforts that affirm the belongingness of people with mental illness and transform social norms to be more respectful of the relational self and more supportive of vulnerable individuals. All of these measures are important to address different aspects of mental illness stigma in various ways.

Mental illness stigma is pervasive and persistent despite great strides made in recent years in increasing awareness of mental health problems and their associated stigma. However, we can reduce and eradicate this stigma through a confluence of actions aimed at eroding some of the core reasons for stigma, including the idealization of social norms that are disrespectful of the relational self and harmful to vulnerable individuals. Mental illness stigma creates many significant harms for those who are stigmatized, but it can be challenged and changed with the right sorts of efforts. Diminishing and eventually eradicating mental illness stigma would help people with mental illness live better lives where they can flourish and thrive in their personal lives and in the community.

References

Abdelmonaem, Yasmin Mohamed Mohamed, Mahmood Ahmed Osman, and Nashwa Ahmed Hussein Abdel Karim. 2024. "Mental Health Stigma and Internship Nursing Students' Attitudes toward Seeking Professional Psychological Help: A Cross-Sectional Study." *BMC Nursing* 23, no. 275: 1–12.

Adu, Joseph, Abe Oudshoom, Kelly Anderson, Carrie Anne Marshall, and Heather Stuart. 2023. "Experiences of Familial Stigma among Individuals Living with Mental Illnesses: A Meta-Synthesis of Qualitative Literature from High-Income Countries." *Journal of Psychiatric and Mental Health Nursing* 30: 208–233.

Adu, Joseph, Abe Oudshoorn, Kelly Anderson, Carrie Anne Marshall, and Heather Stuart. 2022. "Social Contact: Next Steps in an Effective Strategy to Mitigate the Stigma of Mental Illness." *Issues in Mental Health Nursing* 43, no. 5: 485–488.

Adu, Joseph, Abram Oudshoorn, Kelly Anderson, et al. 2021. "Policies and Interventions to Reduce Familial Mental Illness Stigma: A Scoping Review of Empirical Literature." *Issues in Mental Health Nursing* 42, no. 12: 1123–1137.

Ahmad, Shaikh I., Bennett L. Leventhal, Brittany N. Nielsen, and Stephen P. Hinshaw. 2020. "Reducing Mental-Illness Stigma via High School Clubs: A Matched-Pair, Cluster-Randomized Trial." *Stigma and Health* 5, no. 2: 230–239.

Albahari, Miri. 2014. "Alief or Belief? A Contextual Approach to Belief Ascription." *Philosophical Studies* 167: 701–720.

Alexander, Laurel A. and Bruce G. Link. 2003. "The Impact of Contact on Stigmatizing Attitudes toward People with Mental Illness." *Journal of Mental Health* 12, no. 3 (June): 271–289.

Aliche, Chinenye Joseph, Chuka Mike Ifeagwazi, Joy Ugo Nwamarah, Franca Obiageli Okechukwu, and Elias Chukwuemeka Mgwu. 2024. "Mediating Roles of Positive Reappraisal and Stigma Resistance in the Relationship between Mindfulness and Quality of Life among Stable Schizophrenia Patients." *Current Psychology* 43: 3124–3133.

Allport, Gordon W. 1958. *The Nature of Prejudice*. Garden City, NY: Doubleday Anchor Books.

Altamirano, Olivia and Amy Weisman de Mamani. 2024. "Emergency Psychiatric Assistance: Black Individuals' Perceptions of Police and Other Factors That

Influence Mental Health Service Seeking." *Professional Psychology: Research and Practice* 55, no. 3: 278–288.

American Psychiatric Association. 2013. *Diagnostic and Statistical Manual of Mental Disorders*, 5th ed. (*DSM-5*). Washington, DC: American Psychiatric.

Anderson, Elizabeth. 2012. "Epistemic Justice as a Virtue of Social Institutions." *Social Epistemology: A Journal of Knowledge, Culture and Policy* 26, no. 2: 163–173.

Anderson, Elizabeth. 2010. *The Imperative of Integration*. Princeton, NJ: Princeton University Press.

Anderson, Joel and Axel Honneth. 2005. "Autonomy, Vulnerability, Recognition, and Justice." In *Autonomy and the Challenges to Liberalism*, edited by John Christman and Joel Anderson, 127–149. Cambridge: Cambridge University Press.

Angell, Beth, Andrea Cooke, and Kelly Kovac. 2005. "First-Person Accounts of Stigma." In *On the Stigma of Mental Illness: Practical Strategies for Research and Social Change*, edited by Patrick W. Corrigan, 69–98. Washington, DC: American Psychological Association.

Angermeyer, Matthias C., Sandra van der Auwera, Mauro G. Carta, and Georg Schomerus. 2017. "Public Attitudes towards Psychiatry and Psychiatric Treatment at the Beginning of the 21st Century: A Systemic Review and Meta-Analysis of Population Surveys." *World Psychiatry* 16, no. 1 (February): 50–61.

Appelbaum, Paul S. 2020. "Violent Acts and Being the Target of Violence among People with Mental Illness – The Data and Their Limits." *JAMA Psychiatry* 77, no. 4 (April): 345–346.

Aristotle. 2011. *Nicomachean Ethics*. Translated by Robert C. Bartlett and Susan D. Collins. Chicago: University of Chicago Press.

Arneson, Richard J. 1989. "Equality and Equal Opportunity for Welfare." *Philosophical Studies: An International Journal for Philosophy in the Analytic Tradition* 56, no. 1 (May): 77–93.

Atkins, Kim. 2006. "Autonomy and Autonomy Competencies: A Practical and Relational Approach." *Nursing Philosophy* 7: 205–215.

Babij, Alexandra D., Sydney Earl, Whitney Becker, et al. 2025. "Mindsets of Mental Illness: What Are the Implications for Stigma?" *Stigma and Health* 10, no. 3: 443–455 (advance online publication). https://dx.doi.org/10.1037/sah0000482.

Balogun-Mwangi, Oyenike, Nicole R. DeTore, and Zlatka Russinova. 2023. "'We Don't Get a Chance to Prove Who We Really Are': A Qualitative Inquiry of Workplace Prejudice and Discrimination among Black Adults with Serious Mental Illness." *Psychiatric Rehabilitation Journal* 46, no. 1: 4–13.

Balzan, Ryan P., Paul H. Delfabbro, Cherrie A. Galletly, and Todd S. Woodward. 2013. "Illusory Correlations and Control across the Psychosis Continuum." *The Journal of Nervous and Mental Disease* 201, no. 4: 319–327.

Barclay, Linda. 2000. "Autonomy and the Social Self." In *Relational Autonomy: Feminist Perspectives on Autonomy, Agency, and the Social Self*, edited by

Catriona Mackenzie and Natalie Stoljar, 52–71. Oxford: Oxford University Press.

Barnes, Elizabeth. 2016. *The Minority Body: A Theory of Disability*. Oxford: Oxford University Press.

Beeghly, Erin. 2021a. "Stereotyping as Discrimination: Why Thoughts Can Be Discriminatory." *Social Epistemology: A Journal of Knowledge, Culture and Policy* 35, no. 6: 547–563.

Beeghly, Erin. 2021b. "What's Wrong with Stereotypes? The Falsity Hypothesis." *Social Theory and Practice* 47, no. 1 (January): 33–61.

Beltzer, Miranda L. 2020. "Explicit-Implicit Discrepancy in Macro-Level Mental Illness Stigma Is Linked to Prevalence and Care." *Journal of Social and Clinical Psychology* 39, no. 8: 675–708.

Benbow, Sarah, Cheryl Forchuk, Helene Berman, Carolyne Gorlick, and Catherine Ward-Griffin. 2019. "Spaces of Exclusion: Safety, Stigma, and Surveillance of Mothers Experiencing Homelessness." *Canadian Journal of Nursing Research* 51, no. 3: 202–213.

Benson, Paul H. 2015. "Stereotype Threat, Social Belonging, and Relational Autonomy." In *Personal Autonomy and Social Oppression: Philosophical Perspectives*, edited by Marina A. L. Oshana, 124–141. New York: Routledge.

Benson, Paul. 2005. "Taking Ownership: Authority and Voice in Autonomous Agency." In *Autonomy and the Challenges to Liberalism*, edited by John Christman and Joel Anderson, 101–126. Cambridge: Cambridge University Press.

Benson, Paul. 1994. "Free Agency and Self-Worth." *The Journal of Philosophy* 91, no. 12 (December): 650–668.

Bizumic, Boris, Beth Gunningham, and Bruce K. Christensen. 2022. "Prejudice towards People with Mental Illness, Schizophrenia, and Depression among Mental Health Professionals and the General Population." *Psychiatry Research* 217: 1–9 (Article 114817). https://doi.org/10.1016/j.psychres.2022.114817.

Blum, Lawrence. 2004. "Stereotypes and Stereotyping: A Moral Analysis." *Philosophical Papers* 33, no. 3: 251–289.

Bobbili, Sireesha Jennifer, Bruna Sordi Carrara, Raquel Helena Hernandez Fernandes, and Carla Aparecida Arena Ventura. 2022. "A Situational Analysis of Primary Health Care Centers in Brazil: Challenges and Opportunities for Addressing Mental Illness and Substance Use-Related Stigma." *Primary Health Care Research and Development* 23: 1–9 (Article e37). https://doi.org/10.1017/S1463423622000251.

Bojanowska, Agnieszka and Łukasz D. Kaczmarek. 2022. "How Healthy and Unhealthy Values Predict Hedonic and Eudaimonic Well-Being: Dissecting Value-Related Beliefs and Behaviours." *Journal of Happiness Studies* 23, no. 1: 211–231.

Booth, David, Simon Francis, Neil Mcivor, Patrick Hinson, and Benjamin Barton. 2014. "Severe Mental Illness & Employment: Cost-Benefit Analysis and Dynamics of Decision Making." *Mental Health and Social Inclusion* 18, no. 4: 215–223.

Bovens, Luc. 1992. "Sour Grapes and Character Planning." *The Journal of Philosophy* 139, no. 2 (February): 57–78.

Boysen, Guy A. 2017. "Exploring the Relation between Masculinity and Mental Illness Stigma Using the Stereotype Content Model and BIAS Map." *The Journal of Social Psychology* 157, no. 1: 98–113.

Boysen, Guy A., Rebecca L. Chicosky, and Erin E. Delmore. 2023. "Dehumanization of Mental Illness and the Stereotype Content Model." *Stigma and Health* 8, no. 2: 150–158.

Boysen, Guy A., Raina A. Isaacs, Lori Tretter, and Sydnie Markowski. 2020. "Evidence for Blatant Dehumanization of Mental Illness and Its Relation to Stigma." *The Journal of Social Psychology* 160, no. 3: 346–356.

Braedley, Susan and Meg Luxton, eds. 2010. *Neoliberalism and Everyday Life*. Montreal and Kingston: McGill-Queen's University Press.

Breakey, Hugh. 2010. "Adaptive Preferences and the Hellenistic Insight." *Australian Journal of Professional and Applied Ethics* 12, no. 1: 29–39.

Brock, Gillian. 2005. "Needs and Global Justice." *Royal Institute of Philosophy Supplement* 57 (December): 51–72.

Buchanan, Allen E. and Dan W. Brock. 1989. *Deciding for Others: The Ethics of Surrogate Decision Making*. Cambridge: Cambridge University Press.

Burguet, Annette and Frederique Girard. 2024. "Stigma of Schizophrenia and Bipolar Disorders: Explicit and Implicit Measures among Mental Health Professionals." *Stigma and Health* 9, no. 3: 288–293 (advance online publication). https://dx.doi.org/10.1037/sah0000456.

Byrne, Louise, Cath Roper, Brenda Happeli, and Kerry Reid-Searl. 2019. "The Stigma of Identifying as Having a Lived Experience Runs Before Me: Challenges for Lived Experience Roles." *Journal of Mental Health* 28, no. 3: 260–266.

Canham, Sarah L., Piper Moore, Karen Custodio, and Harvey Bosma. 2022. "Homeism: Naming the Stigmatization and Discrimination of Persons Experiencing Homelessness." *Housing, Theory and Society* 39, no. 5: 507–523.

Canton, Emily, Darren Hedley, and Jennifer R. Spoor. 2023. "The Stereotype Content Model and Disabilities." *The Journal of Social Psychology* 163, no. 4: 480–500.

Carel, Havi. 2019. *Illness: The Cry of the Flesh*, 3rd ed. London: Routledge.

Carel, Havi and Ian James Kidd. 2017. "Epistemic Injustice in Medicine and Healthcare." In *The Routledge Handbook of Epistemic Injustice*, edited by Ian James Kidd, Jose Medina, and Gaile, Pohlhaus Jr., 336–346. London: Routledge.

Carel, Havi and Ian James Kidd. 2014. "Epistemic Injustice in Healthcare: A Philosophical Analysis." *Medicine, Heath Care, and Philosophy* 17: 529–540.

Chan, Randolph C. H., Winnie W. S. Mak, and Maggie Y. Y. Lam. 2018. "Self-Stigma and Empowerment as Mediating Mechanisms between Ingroup Perceptions and Recovery among People with Mental Illness." *Stigma and Health* 3, no. 3: 283–293.

Christman, John. 2009. *The Politics of Persons: Individual Autonomy and Sociohistorical Selves*. Cambridge: Cambridge University Press.

Christman, John. 2004. "Relational Autonomy, Liberal Individualism, and the Social Constitution of Selves." *Philosophical Studies* 117: 143–164.

Cohen, Ariel. 1999. *Think Generic: The Meaning and Use of Generic Sentences*. Stanford, CA: Stanford University Center for the Study of Language and Information.

Colburn, Ben. 2011. "Autonomy and Adaptive Preferences." *Utilitas* 23, no. 1 (March): 52–71.

Collins, Patricia Hill. 2000. *Black Feminist Thought: Knowledge, Consciousness, and the Politics of Empowerment*, 2nd ed. New York: Routledge.

Conklin, Teresa M. 2021. "Mental Illness Stigma: Strategies to Address a Barrier to Care." *Women's Healthcare: A Clinical Journal for NPs* 9, no. 2 (April): 16–20.

Conner, Kyaien O., Daniel K. Abusuampeh, Kristin Kosyluk, et al. 2024. "Mitigating the Stigma of Mental Illness: The Impact of Story-Telling in the Black Community." *International Journal of Environmental Research and Public Health* 21, no. 11: 1–13 (Article 1473). https://doi.org/10.3390/ijerph21111473.

Copp, David. 1992. "The Right to an Adequate Standard of Living: Justice, Autonomy, and the Basic Needs." *Social Philosophy & Policy* 9, no. 1: 231–261.

Corrigan, Patrick. 2018. "Beware the Educational Fix: Limitations of Efforts to Promote Mental Health Literacy." *Psychiatric Services* 69, no. 4 (April): 469–471.

Corrigan, Patrick W. and Amy E. Cooper. 2005. "Mental Illness and Dangerousness: Fact or Misperception, and Implications for Stigma." In *On the Stigma of Mental Illness: Practical Strategies for Research and Social Change*, edited by Patrick W. Corrigan, 165-179. Washington, DC: American Psychological Association.

Corrigan, Patrick W., Annette Backs Edwards, Amy Green, Sarah Lickey Diwan, and David L. Penn. 2001. "Prejudice, Social Distance, and Familiarity with Mental Illness." *Schizophrenia Bulletin* 27, no. 2: 219–225.

Corrigan, Patrick W. and Petra Kleinlein. 2005. "The Impact of Mental Illness Stigma." In *On the Stigma of Mental Illness: Practical Strategies for Research and Social Change*, edited by Patrick W. Corrigan, 11–44. Washington, DC: American Psychological Association.

Corrigan, Patrick W. and Katherine Nieweglowski. 2021. "Difference as an Indicator of the Self-Stigma of Mental Illness." *Journal of Mental Health* 30, no. 4: 417–423.

Corrigan, Patrick W. and Amy C. Watson. 2004. "At Issue: Stop the Stigma: Call Mental Illness a Brain Disease." *Schizophrenia Bulletin* 30, no. 3: 477–479.

Courtney, Kellie, Brandon Dial, and Philip T. Yanos. 2024. "'Suck It Up, Go Play': Mental Health Stigma in College Coaches and Their Use of Mental Illness Microaggressions." *Journal of American College Health* 114, no. 9 (September): 1–10. https://doi.org/10.1080/07448481.2024.2400101.

Courtwright, David T. 2002. *Forces of Habit: Drugs and the Making of the Modern World*. Cambridge, MA: Harvard University Press.

Cuddy, Amy J. C., Susan T. Fiske, and Peter Glick. 2002. "A Model of (Often Mixed) Stereotype Content: Competence and Warmth Respectively Follow from Perceived Status and Competition." *Journal of Personality and Social Psychology* 82, no. 6: 878–902.

Culhane, Dennis P., Stephen Metraux, and Trevor Hadley. 2002. "Public Service Reductions Associated with Placement of Homeless Persons with Severe Mental Illness in Supportive Housing." *Housing Policy Debate* 13, no. 1: 107–163.

Daniels, Norman. 2009."Is There a Right to Health Care and, if So, What Does It Encompass?" In *A Companion to Bioethics*, 2nd ed., edited by Helga Kuhse and Peter Singer, 362–372. Malden, MA: Wiley-Blackwell.

Daniels, Norman. 2008. *Just Health: Meeting Health Needs Fairly*. Cambridge: Cambridge University Press, 313–332.

Darwall, Stephen L. 1995. "Two Kinds of Respect." In *Dignity, Character, and Self-Respect*, edited by Robin S. Dillon, 181–197. New York: Routledge.

Davidson, Larry. 2003. *Living Outside Mental Illness: Qualitative Studies of Recovery in Schizophrenia*. New York: New York University Press.

Davy, Laura. 2019. "Between an Ethic of Care and an Ethic of Autonomy: Negotiating Relational Autonomy, Disability, and Dependency." *Angelaki: Journal of the Theoretical Humanities* 24, no. 3: 101–114.

Dean, Jodi. 2008. "Enjoying Neoliberalism." *Cultural Politics* 4, no. 1: 47–72.

De Font-Reaulx, Paul. 2017. "What Makes Discrimination Wrong?" *Journal of Practical Ethics* 5, no. 2: 105–113.

Defourny, Ploon, Nienke van Sambeek, Hester van de Bovenkamp, Floortje Scheepers, and Marjolijn Heerings. 2024. "Dynamics of Recovery in Psychosis, Stigmatization, and Microaggressions in Mental Healthcare: A Qualitative Study of Service Users' Narratives." *BMC Psychiatry* 24: 1–13 (Article 825). https://doi.org/10.1186/s12888-024-06262-5.

Deng, Yali and Theda Rose. 2025. "Examining Dimensions of Help-Seeking Intentions and Associations with Personal Mental Illness Stigma among Racial and Ethnic Minority Immigrants." *Journal of Community Psychology* 53, no. 2 (October): 1–16. https://doi.org/10.1002/jcop.23154.

Desombre, Caroline, Mickaël Jury, Céline Bagès, and Célénie Brasselet. 2019. "The Distinct Effect of Multiple Sources of Stereotype Threat." *The Journal of Social Psychology* 159, no. 5: 628–641.

Dillinger, Rachel L. 2021. "Addressing the Stigma Surrounding Serious Mental Illness in Adolescents: A Brief Intervention." *Psychiatric Quarterly* 92, no. 1: 161–167.

Dinos, Sokratis, Scott Stevens, Marc Serfaty, Scott Weich, and Michael King. 2004. "Stigma: The Feelings and Experiences of 46 People with Mental Illness (Qualitative Study)." *British Journal of Psychiatry* 184: 176–181.

Dobson, Keith S. and Savannah Rose. 2022. "'Myths and Facts' Campaigns Are at Best Ineffective and May Increase Mental Illness Stigma." *Stigma and Health* 7, no. 1: 27–34.

Donnelly, Colleen. 2017. "Public Attitudes toward Disclosing Mental Health Conditions." *Social Work in Mental Health* 15, no. 5: 588–599.

Donovan, Robert J., Catherine F. Drane, Ziggi I. Santini, and Geoffrey Jalleh. 2024. "Impact on Help-Seeking Behaviours of a Campaign Perceived to Decrease Stigma and Increase Openness around Mental Health." *Health Promotion Journal of Australia* 35, no. 4 (October): 1378–1385. https://doi.org/10.1002.hpja.859.

Dorsey, Dale. 2017. "Adaptive Preferences Are a Red Herring." *Journal of the American Philosophical Association* 3, no. 4 (Winter): 465–484.

Dotson, Kristie. 2011. "Tracking Epistemic Violence, Tracking Practices of Silencing." *Hypatia* 26, no. 2: 236–257.

Dotson, Kristie. 2008. "In Search of Tanzania: Are Effective Epistemic Practices Sufficient for Just Epistemic Practices?" *The Southern Journal of Philosophy* 46, 52–64.

Drapalski, Amy L., Natasha Tonge, Anjana Muralidharan, Clayton H. Brown, and Alicia Lucksted. 2023. "Even Mild Internalized Stigma Merits Attention among Adults with Serious Mental Illness." *Psychological Services* 20, no. S2: 166–174.

Dunt, David Robert, Susan Elizabeth Day, Laura Collister, et al. 2022. "Evaluation of a Housing First Programme for People from the Public Mental Health Sector with Severe and Persistent Mental Illnesses and Precarious Housing: Housing, Health and Service Use Outcomes." *Australian & New Zealand Journal of Psychiatry* 56, no. 3: 281–291.

DuPont-Reyes, Melissa J., Alice P. Villatoro, Jared Datzman, et al. 2023. "Inequities Gone or Enduring? Evaluating the Effects of a School-Based Antistigma Intervention on Race/Ethnic and Gender Intersectional Disparities in Mental Illness Stigma." *Stigma and Health* 8, no. 3: 381–392.

Dworkin, Gerald. 1988. *The Theory and Practice of Autonomy*. Cambridge: Cambridge University Press.

Dworkin, Ronald. 1981a. "What Is Equality? Part 1: Equality of Welfare." *Philosophy & Public Affairs* 10, no. 3 (Summer): 185–246.

Dworkin, Ronald. 1981b. "What Is Equality? Part 2: Equality of Resources." *Philosophy & Public Affairs* 10, no. 4 (Autumn): 283–345.

Edwards, Craig. 2010. "Beyond Mental Competence." *Journal of Applied Philosophy* 27, no. 3: 273–289.

Eftekhari, Seena. 2021. "The Irrationality of Adaptive Preferences: A Psychological and Semantic Account." *Utilitas* 33: 68–84.

Ege, Samantha M. and Daniel G. Lannin. 2023. "Deciding to Disclose: The Role of Identity When 'Coming Out Proud.'" *Stigma and Health* 8, no. 2: 205–211.

Elder, Brent C., Carrie E. Rood, and Michelle L. Damiani. 2018. "Writing Strength-Based IEPs for Students with Disabilities in Inclusive Classrooms." *International Journal of Whole Schooling* 14, no. 1: 116–155.

Elliott, Marta and Michael J. Doane. 2015. "Stigma Management of Mental Illness: Effects of Concealment, Discrimination, and Identification on Well-Being." *Self and Identity* 14, no. 6: 654–674.

Ells, Carolyn, Matthew R. Hunt, and Jane Chamber-Evans. 2011. "Relational Autonomy as an Essential Component of Patient-Centered Care." *International Journal of Feminist Approaches to Bioethics* 4, no. 2 (Fall): 79–101.

Elster, Jon. 1983. *Sour Grapes: Studies in the Subversion of Rationality*. Cambridge: Cambridge University Press.

Eminson, Mary. 2004. "Personal Responses to a Lack of Shared Perception." In *Every Family in the Land: Understanding Prejudice and Discrimination Against People with Mental Illness*, revised edition, edited by Arthur H. Crisp, 123–128. London: The Royal Society of Medicine Press.

Farrelly, Simone, Sarah Clement, Jheanell Gabbidon, et al. 2014. "Anticipated and Experienced Discrimination amongst People with Schizophrenia, Bipolar Disorder and Major Depressive Disorder: A Cross Sectional Study." *BMC Psychiatry* 14, no. 157: 1–17. https://doi.org/10.1186/1471-244X-14-157.

Feinberg, Joel. 1986. *Harm to Self: The Moral Limits of the Criminal Law, Volume 2*. Oxford: Oxford University Press.

Feinberg, Joel. 1984. *Harm to Others: The Moral Limits of the Criminal Law, Volume 1*. Oxford: Oxford University Press.

Feinberg, Joel. 1983. "Legal Paternalism." In *Paternalism*, edited by Rolf Sartorius, 3–18. Minneapolis: University of Minnesota Press.

Filia, Kate, Henry Jackson, Sue Cotton, and Eoin Killackey. 2019. "Understanding What It Means to Be Socially Included for People with a Lived Experience of Mental Illness." *International Journal of Social Psychiatry* 65, no. 5: 413–424.

Firmin, Ruth L., Lauren Luther, Paul H. Lysaker, et al. 2018. "Stigma Resistance at the Personal, Peer, and Public Levels: A New Conceptual Model." *Stigma and Health* 2, no. 3 (2017): 182–194.

Fiske, Susan T. 2018. "Stereotype Content: Warmth and Competence Endure." *Current Directions in Psychological Science* 27, no. 2: 67–73.

Fleras, Augie. 2016. "Theorizing Micro-Aggressions as Racism 3.0: Shifting the Discourse." *Canadian Ethnic Studies* 48, no. 2: 1–19.

Follmer, Kayla B. and Kisha S. Jones. 2017. "Stereotype Content and Social Distancing from Employees with Mental Illness: The Moderating Roles of Gender and Social Dominance Orientation." *Journal of Applied Social Psychology* 47, no. 9: 492–504.

Foster, Stephen D., Holger B. Elischberger, and Eric D. Hill. 2018. "Examining the Link between Socioeconomic Status and Mental Illness Prejudice: The Roles of Knowledge about Mental Illness and Empathy." *Stigma and Health* 3, no. 2: 139–151.

Foster, Stephen and Molly O'Mealey. 2022. "Socioeconomic Status and Mental Illness Stigma: The Impact of Mental Illness Controllability Attributions and Personal Responsibility Judgments." *Journal of Mental Health* 31, no. 1: 58–65.

Fox, Annie B. and Valerie A. Earnshaw. 2023. "The Relationship between Mental Illness Stigma and Self-Labeling." *Psychiatric Rehabilitation Journal* 46, no. 2 (December): 1–10. https://dx.doi.org/10.1037/prj0000552.

Fox, Annie B., Dawne Vogt, Jennifer E. Boyd, et al. 2021. "Mental Illness, Problem, Disorder, Distress: Does Terminology Matter When Measuring Stigma?" *Stigma and Health* 6, no. 4: 419–429.

Frankfurt, Harry G. 2003. "Freedom of the Will and the Concept of a Person." In *Free Will*, 2nd ed., edited by Gary Watson, 322–336. Oxford: Oxford University Press.

Freeman, Lauren. 2017. "Embodied Harm: A Phenomenological Engagement with Stereotype Threat." *Human Studies* 40: 637–662.

Fricker, Miranda. 2007. *Epistemic Injustice: Power and the Ethics of Knowing*. Oxford: Oxford University Press.

Friedlaender, Christina. 2018. "On Microaggressions: Cumulative Harm and Individual Responsibility." *Hypatia* 33, no. 1: 5–21.

Friedman, Marilyn. 2003. *Autonomy, Gender, Politics*. Oxford: Oxford University Press.

Fripp, Jessica A. and Ryan G. Carlson. 2017. "Exploring the Influence of Attitude and Stigma on Participation of African American and Latino Populations in Mental Health Services." *Multicultural Counseling and Development* 45 (April): 80–94.

Fung, Kenneth Po-Lun, Jenny Jing-Wen Liu, and Josephine Pui-Hing Wong. 2022. "Exploring Mechanisms of Mental Illness Stigma Reduction in Asian Canadian Men." *The Canadian Journal of Psychiatry* 67, no. 6: 490–498.

Gangi, Cynthia. 2021. "On the Gravity of Mental Illness Stigma." *Philosophy, Psychiatry, & Psychology* 28, no. 4 (December): 385–395.

Garcia-Lorenzo, Lucia, Lucia Sell-Trujillo, and Paul Donnelly. 2022. "Responding to Stigmatization: How to Resist and Overcome the Stigma of Unemployment." *Organization Studies* 43, no. 10: 1629–1650.

Gelder, Michael. 2004. "The Royal College of Psychiatrists' Survey of Public Opinions about Mentally Ill People." In *Every Family in the Land: Understanding Prejudice and Discrimination Against People with Mental Illness*, revised edition, edited by Arthur H. Crisp, 21–25. London: The Royal Society of Medicine Press

Gendler, Tamar Szabó. 2008. "Alief in Action (and Reaction)." *Mind & Language* 23, no. 5 (November): 552–585.

Gergel, Tania Louise. 2014. "Too Similar, Too Different: The Paradoxical Dualism of Psychiatric Stigma." *Psychiatric Bulletin* 38, no. 4: 148–151.

Gibson, Arianna M., Brittany L. Lindsay, Andrew C. H. Szeto. 2025. "Mental Illness Stigma and Microaggressions: An Experimental Study of Familiarity and Relationship Quality." *Stigma and Health* 10, no. 3: 1–11. Advance Online Publication, https://dx.doi.org/10.1037/sah0000471.

Gladman, Beverley and Geoff Waghorn. 2016. "Personal Experiences of People with Serious Mental Illness When Seeking Obtaining and Maintaining Competitive Employment in Queensland, Australia." *Work* 53: 835–843.

Glannon, Walter. 2008. "The Blessing and Burden of Biological Psychiatry." *Journal of Ethics in Mental Health* 3, no. 2: 1–4.

Goffman, Erving. 1963. *Stigma: Notes on the Management of Spoiled Identity.* Englewood Cliffs, NJ: Prentice-Hall.

Gonzales, Lauren, Kristin C. Davidoff, Joseph S. DeLuca, and Philip T. Yanos. 2015. "The Mental Illness Microaggressions Scale-Perpetrator Version (MIMS-P): Reliability and Validity." *Psychiatry Research* 229, nos. 1–2: 120–125.

González-Torres, Miguel Angel, Rodrigo Oraa, Maialen Arístegui, Aranzazu Fernández-Rivas, and Jose Guimon. 2007. "Stigma and Discrimination towards People with Schizophrenia and Their Family Members: A Qualitative Study with Focus Groups." *Social Psychiatry and Psychiatric Epidemiology* 42: 14–23.

Görzig, Anke and Lauren Nicole Ryan. 2022. "The Different Faces of Mental Illness Stigma: Systematic Variation of Stereotypes, Prejudice and Discrimination by Type of Illness." *Scandinavian Journal of Psychology* 63, no. 5 (October): 545–554.

Gosselin, Abigail. 2024. *Mental Health Resilience: The Social Context of Coping with Mental Illness.* Albany, NY: State University of New York Press.

Gosselin, Abigail. 2018. "Mental Illness Stigma and Epistemic Credibility." *Social Philosophy Today* 34: 77–94.

Hacking, Ian. 1999. *The Social Construction of What?* Cambridge, MA: Harvard University Press.

Halpern, Jodi. 2001. *From Detached Concern to Empathy: Humanizing Medical Practice.* Oxford: Oxford University Press.

Hampton, Nan Zhang and Seneca Sharp. 2014. "Internal Motivation to Respond without Prejudice as a Mediator of Gender-Attitudes toward Mental Illness." *Journal of Rehabilitation* 80, no. 3: 30–39.

Harper, David and Ewen Speed. 2012. "Uncovering Recovery: The Resistible Rise of Recovery and Resilience." *Studies in Social Justice* 6, no. 1: 9–25.

Harvey, Jean. 2007. "Moral Solidarity and Empathetic Understanding: The Moral Value and Scope of the Relationship." *Journal of Social Philosophy* 38, no. 1: 22–37.

Haslam, Nick. 2006. "Dehumanization: An Integrative Review." *Personality and Social Psychology Review* 10, no. 3: 252–264.

Haslam, Nick and Erlend P. Kvaale. 2015. "Biogenetic Explanations of Mental Disorder: The Mixed-Blessings Model." *Current Directions in Psychological Science* 24, no. 5: 399–404.

Haugland, Siri Håvås, Alain Topor, and Jan Georg Friesinger. 2024. "1 Plus 1 Is More Than 2: Mental Health Problems, Financial Difficulties, and Social Exclusion in a Cross-Sectional Study of 28,047 General-Population Adults." *BMC Public Health* 24: 1–9 (Article 1148). https://doi.org/10.1186/s12889-024-18555-1.

Hehman, Eric and Rebecca Neel. 2024. "Prejudice Model 1.0: A Predictive Model of Prejudice." *Psychological Review* 131, no. 5: 1235–1265.

Hejizadeh, Alireza, Homayoun Amini, Mahdiyeh Heydari, and Fatemeh Rajabi. 2024. "How to Combat Stigma Surrounding Mental Health Disorders:

A Scoping Review of the Experiences of Different Stakeholders." *BMC Psychiatry* 24, no. 782: 1–14. https://doi.org/10.1186/s12888-024-06220-1.

Hellman, Deborah. 2008. *When Is Discrimination Wrong?* Cambridge, MA: Harvard University Press.

Helu-Brown, Paula, Concepcion Barrio, and Mercedes Hernandez. 2023. "Stigma and the Intersection of Ethnicity, Gender, Romantic Relationships, and Serious Mental Illness: Perspectives from Latinxs with Schizophrenia and Their Family Members." *Stigma and Health* 8, no. 3: 344–354.

Henderson, Claire, Jo Neblett, Hannah Parke, et al. 2014. "Mental Health-Related Stigma in Health Care and Mental Health-Care Settings." *Lancet Psychiatry* 1, no. 6 (November): 467–482.

Heney, Diana B. 2022. "Solving for Stigma in Mental Health Care." *Journal of Evaluation in Clinical Practice* 28: 883–889.

Hennekam, Sophie, Kayla Follmer, and Joy E. Beatty. 2021. "The Paradox of Mental Illness and Employment: A Person-Job Fit Lens." *The International Journal of Human Resource Management* 32, no. 15: 3244–3271.

Henry, Julie D., Courtney von Hippel, and Lisa Shapiro. 2010. "Stereotype Threat Contributes to Social Difficulties in People with Schizophrenia." *British Journal of Clinical Psychology* 49, no. 1: 31–41.

Henwood, Benjamin F., Jason Matejkowski, Ana Stefancic, and Jonathan M. Lukens. 2014. "Quality of Life after *Housing First* for Adults with Serious Mental Illness Who Have Experienced Chronic Homelessness." *Psychiatry Research* 220: 549–555.

Hipes, Crosby. 2016. "Stereotypes, Power, and the Stigmatization of Workers with Mental Illness." *Conference Papers—American Sociological Association*: 1–34.

Hoffman, Ginger A. and Jennifer L. Hansen. 2017. "Prozac or Prosaic Diaries?: The Gendering of Psychiatric Disability in Depression Memoirs." *Philosophy, Psychiatry & Psychology* 24, no. 4 (December): 285–298.

Holland, Kate. 2012. "The Unintended Consequences of Campaigns Designed to Challenge Stigmatising Representations of Mental Illness in the Media." *Social Semiotics* 22, no. 3 (June): 217–236.

Holmes, Hannah and Gemma Burgess. 2021. "Homelessness Prevention through One-to-One Coaching: The Relationship between Coaching, Class Stigma, and Self-Esteem." *Housing, Theory and Society* 38, no. 5: 580–596.

Holroyd, Jules. 2012. "Responsibility for Implicit Bias." *Journal of Social Philosophy* 43, no. 3 (Fall): 274–306.

Holroyd, Jules, Robin Scaife, and Tom Stafford. 2017. "What Is Implicit Bias?" *Philosophy Compass* 12: e12437.

Hookway, Christopher. 2010. "Some Varieties of Epistemic Injustice: Reflections on Fricker." *Episteme* 7, no. 2: 151–163.

Horta, Oscar. 2015. "Does Discrimination Require Disadvantage?" *Moral Philosophy and Politics* 2, no. 2: 277–297.

Howarth, Caroline. 2006. "Race as Stigma: Positioning the Stigmatized as Agents, Not Objects." *Journal of Community & Applied Social Psychology* 16: 442–451.

Janda, Ami C. 2008. "Keeping a Productive Labor Market: Crafting Recognition and Rights for Mentally Ill Workers." *Hamline Journal of Public Law & Policy* 30, no. 1: 403–445.
Johnson-Kwochka, Annalee, Gary R. Bond, Deborah R. Becker, Robert E. Drake, and Mary Ann Greene. 2017. "Prevalence and Quality of Individual Placement and Support (IPS) Supported Employment in the United States." *Administration and Policy in Mental Health and Mental Health Services Research* 44: 311–319.
Jones, Amanda M. 2011. "Disclosure of Mental Illness in the Workplace: A Literature Review." *American Journal of Psychiatric Rehabilitation* 14: 212–229.
Kachur, Elizabeth and Thomas Harter. 2024. "Wired for Stereotyping and Biases—Is There a Professional Way Out of Prejudicial Behaviors and Discrimination?" *Journal of Communication in Healthcare* 17, no. 4: 355–359.
Kant, Immanuel. 1993. *Grounding for the Metaphysics of Morals*, 3rd ed. Translated by James W. Ellington. Indianapolis, IN: Hackett.
Kao, Shiao Yan Sharon, Amanda C. M. Li, Winnie W. S. Mak, and Rebecca Y. M. Cheung. 2024. "The Efficacy of Acceptance and Commitment Therapy on Self-Stigma Reduction among People with Mental Illness: A Quasi-Experimental Design." *Stigma and Health* 9, no. 2: 1–11 (advance online publication). https://dx.doi.org/10.1037/sah0000474.
Katsekas, Bette and Diane Lemay. 1996. "Dynamics of Prejudice: A Proposed Definition and the Counselor's Role in the Evolution of Attitudes." *ERIC (Education Resources Information Center)*: 1–9. Accessed January 6, 2025. https://eric.ed.gov/?id=ED418369.
Keller, Jean. 1997. "Autonomy, Relationality, and Feminist Ethics." *Hypatia* 12, no. 2 (Spring): 152–164.
Kelly, Erin. 2004. "Human Rights as Foreign Policy Imperatives." In *The Ethics of Assistance: Morality and the Distant Needy*, edited by Deen K. Chatterjee, 177–192. Cambridge: Cambridge University Press.
Kenny, Amanda and Boris Bizumic. 2016. "Learn and ACT: Changing Prejudice towards People with Mental Illness Using Stigma Reduction Interventions." *Journal of Contextual Behavioral Science* 5: 178–185.
Kerman, Nick, John Sylvestre, Tim Aubry, and Jino Distasio. 2018. "The Effects of Housing Stability on Service Use among Homeless Adult with Mental Illness in a Randomized Controlled Trial of Housing First." *BMC Health Services Research* 18, no. 190: 1–14. https://doi.org/10.1186/s12913-018-3028-7.
Khader, Serene J. 2020. "The Feminist Case against Relational Autonomy." *Journal of Moral Philosophy* 17: 499–526.
Kidd, Ian James. 2019. "Pathophobia, Vices, and Illness." *International Journal of Philosophical Studies* 27, no. 2: 286–306.
Kidd, Ian James and Havi Carel. 2021. "The Predicament of Patients." *Royal Institute of Philosophy Supplement* 89: 65–84.
Kidd, Ian James and Havi Carel. 2018. "Healthcare Practice, Epistemic Injustice, and Naturalism." *Royal Institute of Philosophy Supplement* 84: 211–233.

Kidd, Ian James and Havi Carel. 2017. "Epistemic Injustice and Illness." *Journal of Applied Philosophy* 34, no. 2 (February): 172–190.

Koenig, Anne M. and Alice H. Eagly. 2014. "Extending Role Congruity Theory of Prejudice to Men and Women with Sex-Typed Mental Illnesses." *Basic and Applied Social Psychology* 36: 70–82.

Kolodny, Niko. 2014. "Rule Over None II: Social Equality and the Justification of Democracy." *Philosophy & Public Affairs* 42, no. 4: 287–336.

Krupa, Terry, Bonnie Kirsh, Lynn Cockburn, and Rebecca Gewurtz. 2009. "Understanding the Stigma of Mental Illness in Employment." *Work* 33: 413–425.

Kuljian, Olivia R. and Zachary P. Hohman. 2023. "Warmth, Competence, and Subtle Dehumanization: Comparing Clustering Patterns of Warmth and Competence with Animalistic and Mechanistic Dehumanization." *British Journal of Social Psychology* 62, no. 1: 181–196.

Kurs, Rena and Alexander Grinshpoon. 2018. "Vulnerability of Individuals with Mental Disorders to Epistemic Injustice in Both Clinical and Social Domains." *Ethics & Behavior* 28, no. 4: 336–346.

Lakeman, Richard. 2010. "Epistemic Injustice and the Mental Health Service User." *International Journal of Mental Health Nursing* 19: 151–153.

Leclair, Marichelle C., Félicia Deveaux, Maurence Roy, et al. 2019. "The Impact of Housing First on Criminal Justice Outcomes among Homeless People with Mental Illness: A Systematic Review." *The Canadian Journal of Psychiatry* 64, no. 8: 525–530.

Lemke, Thomas. 2001. "'The Birth of Bio-Politics': Michel Foucault's Lecture at the Collège de France on Neo-Liberal Governmentality." *Economy and Society* 30, no. 2: 190–207.

Levy, Neil. 2017. "Implicit Bias and Moral Responsibility: Probing the Data." *Philosophy and Phenomenological Research* 94, no. 1 (January): 3–26.

Levy, Neil. 2015. "Neither Fish nor Fowl: Implicit Attitudes as Patchy Endorsements." *Noûs* 49, no. 4: 800–823.

Liamputtong, Pranee and Zoe Sanipreeya Rice. 2022. "Stigma, Discrimination, and Social Exclusion." In *Handbook of Social Inclusion: Research and Practices in Health and Social Sciences*, edited by Pranee Liamputtong, 113–128. Cham, Switzerland: Springer.

Link, Bruce G. and Jo C. Phelan. 2001. "Conceptualizing Stigma." *Annual Review of Sociology* 27: 363–385.

Lippert-Rasmussen, Kasper. 2014a. *Born Free and Equal? A Philosophical Inquiry into the Nature of Discrimination*. Oxford: Oxford University Press.

Lippert-Rasmussen, Kasper. 2014b. "Indirect Discrimination Is Not Necessarily Unjust." *Journal of Practical Ethics* 2, no. 2: 33–57.

Loubière, S., C. Lemoine, M. Boucekine, et al. 2022. "Housing First for Homeless People with Severe Mental Illness: Extended 4-Year Follow-Up and Analysis of Recovery and Housing Stability from the Randomized *Un Chez Soi d'Abord* Trial." *Epidemiology and Psychiatric Sciences* 31: e14. https://doi.org/10.1017/S2045796022000026.

Louden, Jennifer Eno, Alondra Avila, Oscar Armando Esparza del Villar, et al. 2023. "Self-Stigma of Mental Illness among Latino People in the U.S.-México Border." *Stigma and Health* 8, no. 1: 21–31.

Lui, Joyce Claresta, Ilyas Sagar-Ouriaghli, and June S. L. Brown. 2024. "Barriers and Facilitators to Help-Seeking for Common Mental Disorders among University Students: A Systematic Review." *Journal of American College Health* 72, no. 8: 2605–2613.

Luong, Linh, James Lachaud, Fiona G. Kouyoumdjian, Stephen W. Hwang, and Cilia Mejia-Lancheros. 2021. "The Impact of a Housing First Intervention and Health-Related Risk Factors on Incarceration among People with Experiences of Homelessness and Mental Illness in Canada." *Canadian Journal of Public Health* 112: 270–279.

Mackenzie, Catriona. 2014. "Three Dimensions of Autonomy: A Relational Analysis." In *Autonomy, Oppression, and Gender*, edited by Andrea Veltman and Mark Piper, 15–41. Oxford: Oxford University Press.

Mackenzie, Catriona. 2008. "Relational Autonomy, Normative Authority and Perfectionism." *Journal of Social Philosophy* 39, no. 4 (Winter): 512–533.

Mackenzie, Catriona and Natalie Stoljar. 2000. "Introduction: Autonomy Refigured." In *Relational Autonomy: Feminist Perspectives on Autonomy, Agency, and the Social Self*, edited by Catriona Mackenzie and Natalie Stoljar, 3–31. Oxford: Oxford University Press.

Maeshima, Lindsey S. and Mike C. Parent. 2022. "Mental Health Stigma and Professional Help-Seeking Behaviors among Asian American and Asian International Students." *Journal of American College Health* 70, no. 6: 1761–1767.

Malnes, Raino. 2015. "Discrimination: Classification and Moral Assessment." *Moral Philosophy and Politics* 2, no. 2: 245–254.

Mann, Caroline E. and Yareli Sosa Antunez. 2024. "Dangerous Intersection: An Experimental Examination of Layered or 'Double Stigma' Toward Latiné Persons Living with Mental Illness." *Stigma and Health*: 1–12. Advance online publication, https://dx.doi.org/10.1037/sah0000565.

Marcussen, Kristen, Mary Gallagher, and Christian Ritter. 2021. "Stigma Resistance and Well-Being in the Context of the Mental Illness Identity." *Journal of Health and Social Behavior* 62, no. 1: 19–36.

Marinucci, Marco, Paolo Riva, Michela Lenzi, et al. 2023. "On the Lowest Rung of the Ladder: How Social Exclusion, Perceived Economic Inequality and Stigma Increase Homeless People's Resignation." *British Journal of Social Psychology* 62, no. 4: 1817–1838.

Markowitz, Fred E. and Jeffrey Syverson. 2021. "Race, Gender, and Homelessness Stigma: Effects of Perceived Blameworthiness and Dangerousness." *Deviant Behavior* 42, no. 7: 919–931.

McAllister, Ashley, Kara Dickson, Mediya Rangi, et al. 2023. "Embedding Interpersonal Stigma Resistance into the Medical Curriculum: A Focus Group Study of Medical Students." *BMC Medical Education* 23, no. 1: 1–12 (Article 686). https://doi.org/10.1186/s12909-023-04512-w.

McCullock, Seth P. and Rachel M. Scrivano. 2023. "The Effectiveness of Mental Illness Stigma-Reduction Interventions: A Systematic Meta-Review of Meta-Analyses." *Clinical Psychology Review* 100: 1–14 (Article 102242).

McKinnon, Rachel. 2014. "Stereotype Threat and Attributional Ambiguity for Trans Women." *Hypatia* 29, no. 4 (Fall): 857–872.

Mejia-Lancheros, Cilia, James Lachaud, Julia Woodhall-Melnik, et al. 2021. "Longitudinal Interrelationships of Mental Health Discrimination and Stigma with Housing and Well-Being Outcomes in Adults with Mental Illness and Recent Experience of Homelessness." *Social Science & Medicine* 268 (January): 1–14 (Article 113463).

Mestdagh, Annelin and Bart Hansen. 2014. "Stigma in Patients with Schizophrenia Receiving Community Mental Health Care: A Review of Qualitative Studies." *Social Psychiatry and Psychiatric Epidemiology* 49: 79–87.

Metzl, Jonathan M. 2009. *The Protest Psychosis: How Schizophrenia Became a Black Disease*. Boston: Beacon Press.

Meyers, Diana Tietjens. 2004. *Being Yourself: Essays on Identity, Action, and Social Life*. Lanham, MD: Rowman & Littlefield.

Mezey, Gillian, Sarah White, Isobel Harrison, et al. 2022. "Modelling Social Exclusion in a Diagnostically-Mixed Sample of People with Severe Mental Illness." *International Journal of Social Psychiatry* 68, no. 2: 420–428.

Middleton, David. 2006. "Three Types of Self-Respect." *Res Publica* 12, no. 1: 59–76.

Mikkola, Mari. 2016. *The Wrong of Injustice: Dehumanization and Its Role in Feminist Philosophy*. Oxford: Oxford University Press.

Mills, Charles W. 2018. "Through a Glass Whitely: Ideal Theory as Epistemic Injustice." *Proceedings & Addresses of the American Philosophical Association* 92 (November): 43–77.

Mills, Charles W. 2005. "'Ideal Theory' as Ideology." *Hypatia* 20, no. 3 (Summer): 165–184.

Misra, Supriya, Valerie W. Jackson, Jeanette Chong, et al. 2021. "Systematic Review of Cultural Aspects of Stigma and Mental Illness among Racial and Ethnic Minority Groups in the United States: Implications for Interventions." *American Journal of Community Psychology* 68: 486–512.

Mitchell, Polly. 2018. "Adaptive Preferences, Adapted Preferences." *Mind* 127, no. 508 (October): 1003–1025.

Moreau, Sophia. 2020. *Faces of Inequality: A Theory of Wrongful Discrimination*. Oxford: Oxford University Press.

Morrow, Marina, Stephanie Bryson, Rodrick Lai, et al. 2020. "Intersectionality as an Analytic Framework for Understanding the Experiences of Mental Health Stigma among Racialized Men." *International Journal of Mental Health and Addiction* 18: 1304–1317.

Na, Jennifer Jiwon, Joanne L. Park, Tuguldur Lkhagva, and Amori Yee Mikami. 2022. "The Efficacy of Interventions on Cognitive, Behavioral, and Affective Public Stigma around Mental Illness: A Systematic Meta-Analytic Review." *Stigma and Health* 7, no. 2: 127–141.

Nanay, Bence. 2021. "Implicit Bias as Mental Imagery." *Journal of the American Philosophical Association* 7, no. 3 (Fall): 329–347.

National Alliance to End Homelessness. 2016. "Housing First." *National Alliance to End Homelessness* (April 20). Accessed April 1, 2022. https://endhomelessness.org/resource/housing-first/?msclkid=a2c7065db20011ec8efb938e6962f5e0.

Nedelsky, Jennifer. 2011. *Law's Relations: A Relational Theory of Self, Autonomy, and Law*. Oxford: Oxford University Press.

Nedelsky, Jennifer. 1989. "Reconceiving Autonomy: Sources, Thoughts and Possibilities." *Yale Journal of Law and Feminism* 1, no. 7: 7–36.

Nirmala, Bergai Parthasarathy, Rania Roy, Vidya Naik, and Pallerla Srikanth. 2020. "Employability of People with Mental Illness and Substance Use Problems: Field Realities." *Journal of Family Medicine and Primary Care* 9: 3405–3410.

O'Connor, Cliodhna, Maryanne Brassil, Sadhbh O'Sullivan, Christina Seery, and Finiki Nearchou. 2022. "How Does Diagnostic Labelling Affect Social Responses to People with Mental Illness? A Systematic Review of Experimental Studies Using Vignette-Based Designs." *Journal of Mental Health* 31, no. 1: 115–130.

O'Hara, Mary. 2014. "Mental Health and Race: The Blight of Dual Discrimination: Black and Minority Ethnic People Living with Mental Illness Face Double Prejudice." *The Guardian* (March 26): A40.

Okamoto, Kristen E. and Brittany L. Peterson. 2022. "Resurrecting and Appending Identities: The Role of Nonprofits in Managing Stigmatized Identity." *Nonprofit and Voluntary Sector Quarterly* 51, no. 2: 389–409.

O'Neill, Onora. 1993. "Justice, Gender, and International Boundaries." In *The Quality of Life*, edited by Martha Nussbaum and Amartya Sen, 303–323. Oxford: Clarendon Press.

O'Neill, Onora. 1987. "Abstraction, Idealization and Ideology in Ethics." In *Moral Philosophy and Contemporary Problems*, edited by John David Gemmill Evans, 55–69. Cambridge: Cambridge University Press.

Ong, Anthony D., Cindy S. Bergeman, and Sy-Miin Chow. 2010. "Positive Emotions as a Basic Building Block of Resilience in Adulthood." In *Handbook of Adult Resilience*, edited by John W. Reich, Alex J. Zautra, and John Stuart Hall, 81–93. New York: Guilford Press.

Onyeukwu, Ugochinyere N. and John J. Donahue. 2024. "Mental Health Treatment Stigma, Maladaptive Personality Trait Domains, and Treatment-Seeking Attitudes and Behaviors." *Journal of Clinical Psychology* 80: 421–436, doi: 10.1002/jclp.23619.

Oshana, Marina. 2007. "Autonomy and the Question of Authenticity." *Social Theory and Practice* 33, no. 3 (July): 411–429.

Oshana, Marina A. L. 2005. "Autonomy and Self-Identity." In *Autonomy and the Challenges to Liberalism*, edited by John Christman and Joel Anderson, 77–97. Cambridge: Cambridge University Press.

Oshana, Marina. 2003. "How Much Should We Value Autonomy?" *Social Philosophy & Policy* 20, no. 2: 99–126.

Oshana, Marina A. L. 1998. "Personal Autonomy and Society." *Journal of Social Philosophy* 29, no. 1 (Spring): 81–102.

Østerud, Kaja Larsen. 2023. "Mental Illness Stigma and Employer Evaluation in Hiring: Stereotypes, Discrimination and the Role of Experience." *Sociology of Health & Illness* 43: 90–108.

Ottati, Victor, Galen V. Bodenhausen, and Leonard S. Newman. 2005. "Social Psychological Models of Mental Illness Stigma." In *On the Stigma of Mental Illness: Practical Strategies for Research and Social Change*, edited by Patrick W. Corrigan, 99–128. Washington, DC: American Psychological Association.

Oudejans, Suzan C. C., M. E. Spits, and Jaapvan Weeghel. 2021. "A Cross-Sectional Survey of Stigma towards People with a Mental Illness in the General Public: The Role of Employment, Domestic Noise Disturbance and Age." *Social Psychiatry and Psychiatric Epidemiology* 56: 1547–1554.

Özcan, Fatma Yener and Burcu Ceylan. 2022. "Lawyers' Beliefs about Mental Illnesses and Their Tendency to Stigmatize People with Mental Illnesses." *Perspectives in Psychiatric Care* 58, no. 4 (October): 1303–1309.

Padgett, Deborah K., Benjamin F. Henwood, and Sam J. Tsemberis. 2016. *Housing First: Ending Homelessness, Transforming Systems, and Changing Lives.* Oxford: Oxford University Press.

Pasnau, Robert. 2022. "Bias and Interpersonal Skepticism." *Noûs* 56, no. 1 (March): 154–175.

Patterson, Michelle, Akm Moniruzzaman, Anita Palepu, et al. 2013. "Housing First Improves Subjective Quality of Life among Homeless Adults with Mental Illness: 12-Month Findings from a Randomized Controlled Trial in Vancouver, British Columbia." *Social Psychiatry & Psychiatric Epidemiology* 48: 1245–1259.

Peet, Andrew. 2017. "Epistemic Injustice in Utterance Interpretation." *Synthese* 194: 3421–3443.

Perlin, Michael L. 2016. "'Infinity Goes Up On Trial': Sanism, Pretextuality, and the Representation of Defendants with Mental Disabilities." *QUT Law Review* 16, no. 3 (December): 106–126.

Perlin, Michael L. 2000. *The Hidden Prejudice: Mental Disability on Trial.* Washington, DC: American Psychological Association.

Perlin, Michael L. 1993. "On Sanism." *SMU Law Review* 46, no. 2: 373–407.

Perlin, Michael L. and Deborah A. Dorfman. 1993. "Sanism, Social Science, and the Development of Mental Disability Law Jurisprudence." *Behavioral Sciences and the Law* 11: 47–66.

Peters, Heather J., Haley N. Schwenk, Zachary R. Ahlstrom, and Lyndzie N. McIalwain. 2017. "Microaggressions: The Experience of Individuals with Mental Illness." *Counselling Psychology Quarterly* 30, no. 1: 86–112.

Phelan, Jo, Bruce G. Link, Robert E. Moore, and Ann Stueve. 1997. "The Stigma of Homelessness: The Impact of the Label 'Homeless' on Attitudes toward Poor Persons." *Social Psychology Quarterly* 60, no. 4 (December): 323–337.

Pilling, Merrick D. 2021. "Sexual Violence and Psychosis: Intersections of Rape Culture, Sanism, and Anti-Black Sanism in Psychiatric Inpatient Chart Documentation." In *Interrogating Psychiatric Narratives of Madness: Documented*

Lives, edited by Andrea Daley and Merrick D. Pilling, 137–163. Cham: Palgrave Macmillan.

Pogge, Thomas. 2008. *World Poverty and Human Rights*, 2nd ed. Cambridge: Polity.

Qin, Sang, Lindsay Sheehan, Eva Yau, et al. 2024. "Adapting and Evaluating a Strategic Disclosure Program to Address Mental Health Stigma among Chinese." *International Journal of Mental Health and Addiction* 22, no. 3: 1029–1046.

Radden, Jennifer. 2000. "Seligman: A Learned Helplessness Model of Depression." In *The Nature of Melancholy: From Aristotle to Kristeva*, edited by Jennifer Radden, 311–316. Oxford: Oxford University Press.

Rangel, Ulrike and Johannes Keller. 2011. "Essentialism Goes Social: Belief in Social Determinism as a Component of Psychological Essentialism." *Journal of Personality and Social Psychology* 100, no. 6: 1056–1078.

Rawls, John. 1971. *A Theory of Justice*. Cambridge, MA: Belknap Press.

Raz, Joseph. 1986. *The Morality of Freedom*. Oxford: Clarendon Press.

Read, John, Nick Haslam, Liz Sayce, and E. Bethan Davies. 2006. "Prejudice and Schizophrenia: A Review of the 'Mental Illness Is an Illness Like Any Other' Approach." *Acta Psychiatrica Scandinavica* 114: 303–318.

Reader, Soran. 2007. *Needs and Moral Necessity*. New York: Routledge.

Rebeiro Gruhl, Karen L., Carol Kauppi, Phyllis Montgomery, and Susan James. 2012. "Painting Everybody with the Same Brush: Employment Discrimination of Persons with Serious Mental Illness in Rural Places." *Rural Mental Health* 36, no. 11: 11–17.

Reich, John W., Alex J. Zautra, and John Stuart Hall. 2010. "Preface." In *Handbook of Adult Resilience*, edited by John W. Reich, Alex J. Zautra, and John Stuart Hall, xi–xv. New York: Guilford Press.

Reiheld, Alison. 2010. "Patient Complains Of . . .: How Medicalization Mediates Power and Justice." *International Journal of Feminist Approaches to Bioethics* 3, no. 1 (Spring): 72–98.

Reilly, Johanna, Iris Ho, and Andrea Williamson. 2022. "A Systematic Review of the Effect of Stigma on the Health of People Experiencing Homelessness." *Health and Social Care in the Community* 30: 2128–2141.

Reinhard, Matthias A., Julia Dewald-Kaufman, Torsten Wüstenberg, et al. 2020. "The Vicious Circle of Social Exclusion and Psychopathology: A Systematic Review of Experimental Ostracism Research in Psychiatric Disorders." *European Archives of Psychiatry and Clinical Neuroscience* 270, no. 5: 521–532.

Rickard, M. 1995. "Sour Grapes, Rational Desires and Objective Consequentialism." *Philosophical Studies: An International Journal for Philosophy in the Analytic Tradition* 80, no. 3 (December): 279–303.

Rieckhof, Sophia, Christian Sander, Sven Speerforck, et al. 2021. "Development and Validity of the Value-Based Stigma Inventory (VASI): A Value-Sensitive Questionnaire for the Assessment of Mental Health Stigma." *BMC Psychiatry* 21, no. 1: 1–11 (Article 570). https://doi.org/10.1186/s12888-021-03427-4.

Riffel, Taylor and Shu-Ping Chen. 2020. "Stigma in Healthcare? Exploring the Knowledge, Attitudes, and Behavioural Responses of Healthcare Professionals and Students toward Individuals with Mental Illnesses." *Psychiatric Quarterly* 91: 1103–1119.

Rippere, Vicky. 1977. "Comments on Seligman's Theory of Helplessness." *Behaviour Research and Therapy* 15, no. 2: 207–209.

Robb, Jayci and Jeff Stone. 2016. "Implicit Bias toward People with Mental Illness: A Systematic Literature Review." *Journal of Rehabilitation* 82, no. 4: 3–13.

Rodwin, Aaron H., Rei Shimizu, Moiyattu Banya, et al. 2025. "Stigma among Historically Marginalized Young Adults with Serious Mental Illnesses: A Mixed Methods Study." *Stigma and Health* 10, no. 1: 1–13 (advance online publication). https://dx.doi.org/10.1037/sah0000454.

Roessler, Beate. 2015. "Autonomy, Self-Knowledge, and Oppression." In *Personal Autonomy and Social Oppression: Philosophical Perspectives*, edited by Marina A. L. Oshana, 68–84. New York: Routledge.

Rüsch, Nicolas, Matthias C. Angermeyer, and Patrick W. Corrigan. 2005. "Mental Illness Stigma: Concepts, Consequences, and Initiatives to Reduce Stigma." *European Psychiatry* 20: 529–539.

Sadler, Melody S., Elizabeth L. Meagor, and Kimberly E. Kaye. 2012. "Stereotypes of Mental Disorders Differ in Competence and Warmth." *Social Science & Medicine* 74, no. 6: 915–922.

Saldua, Mark. 2023. "Addressing Social Determinants of Health among Individuals Experiencing Homelessness." *SAMHSA (Substance Abuse and Mental Health Services Administration)*, (November 15). Accessed January 1, 2025. www.samhsa.gov/blog/addressing-social-determinants-health-among-individuals-experiencing-homelessness.

Salk, Rachel H., Janet S. Hyde, and Lyn Y. Abramson. 2017. "Gender Differences in Representative National Samples: Meta-Analyses of Diagnoses and Symptoms." *Psychological Bulletin* 143, no. 8: 783–822.

Sanati, Abdi and Michalis Kyratsous. 2015. "Epistemic Injustice in Assessment of Delusions." *Journal of Evaluation in Clinical Practice* 21: 479–485.

Saul, Jennifer. 2013. "Skepticism and Implicit Bias." *Disputatio: International Journal of Philosophy* 5, no. 37: 243–263.

Schindler, Victoria P. and Camille Sauerwald. 2013. "Outcomes of a 4-Year Program with Higher Education and Employment Goals for Individuals Diagnosed with Mental Illness." *Work* 46: 325–336.

Schomerus, Georg and Matthias Claus Angermeyer. 2021. "Blind Spots in Stigma Research? Broadening Our Perspective on Mental Illness Stigma by Exploring 'What Matters Most' in Modern Western Societies." *Epidemiology and Psychiatric Sciences* 30, no. e26: 1–6. https://doi.org/10.1017/S2045796021000111.

Schreiter, Stefanie, Sven Speerforck, Georg Schomerus, and Stefan Gutwinski. 2021. "Homelessness: Care for the Most Vulnerable: A Narrative Review of Risk Factors, Health Needs, Stigma, and Intervention Strategies." *Current Opinion in Psychiatry* 34, no. 4 (July): 400–404.

Schulze, Beate and Matthias C. Angermeyer. 2003. "Subjective Experiences of Stigma: A Focus Group Study of Schizophrenic Patients, Their Relatives and Mental Health Professionals." *Social Science & Medicine* 56: 299–312.

Schwartz, Shalom H. 2012. "An Overview of the Schwartz Theory of Basic Values." *Online Readings in Psychology and Culture* 2, no. 1: 1–20 (Chapter 11). https://doi.org/10.9707/2307-0919.1116.

Schwitzgebel, Eric. 2010. "Acting Contrary to Our Professed Beliefs or the Gulf between Occurrent Judgment and Dispositional Belief." *Pacific Philosophical Quarterly* 91: 531–553.

Segev, Re'em. 2014. "Making Sense of Discrimination." *Ratio Juris: An International Journal of Jurisprudence and Philosophy of Law* 27, no. 1 (March): 47–78.

Seligman, Martin E. P. 1975. *Helplessness: On Depression, Development, and Death.* New York: W. H. Freeman.

Seo, Mikyung, Minhwa Lee, and Jinhyang Lee. 2024. "Effects of Group Identification and Stigma Resistance on Affiliate Stigma and Empowerment among Families of People with Mental Illness." *Asia Pacific Journal of Social Work and Development* 34, no. 4: 366–379.

Shannon, Jordan, Derek X. Seward, Yanhong Liu, and Melissa Luke. 2022. "Stigma, Help-Seeking, and Counseling with African American Male College Students." *Journal of Counseling and Development* 100, no. 4: 421–432.

Shearer, Amy L., Elizabeth Roth, Mathew S. Cefalu, et al. 2021. "Contact with Persons with Mental Illness and Willingness to Live Next Door To Them: Two Waves of a California Survey of Adults." *Psychiatric Services* 72, no. 1 (January): 23–30.

Sherwin, Susan. 2000. "A Relational Approach to Autonomy in Health Care." In *Readings in Health Care Ethics*, edited by Elisabeth Boetzkes and Wilfrid J. Waluchow, 69–87. Ontario: Broadview Press.

Silván-Ferrero, Prado, Pablo F. Holgado, José Jiménez, and Daniel Pérez-Garin. 2022. "Benefits of Employment in People with Mental Illness: Differential Mediating Effects of Internalized Stigma on Self-Esteem." *Journal of Community & Applied Social Psychology* 32: 119–134.

Smith, David Livingstone. 2016. "Paradoxes of Dehumanization." *Social Theory and Practice* 42, no. 2 (April): 416–443.

Smith, David Livingstone. 2014. "Dehumanization, Essentialism, and Moral Psychology." *Philosophy Compass* 9, no. 11: 814–824.

Smith, Jacqueline M., Stephanie Knaak, Jennifer Smith, et al. 2024. "An Exploration of Mental Health-Related Stigma in an Emergency Setting." *International Journal of Mental Health and Addiction* 23, no. 3: 1–23. https://doi.org/10.1007/s11469-023-01233-5.

Sönmez, Burçak and Kübra Meltem Karaoğlu. 2023. "Contents of Stereotypes toward Mental Illness." *Current Psychology* 42, no. 30: 26545–26554.

Steele, Claude M. 1997. "A Threat in the Air: How Stereotypes Shape Intellectual Identity and Performance." *American Psychologist* 52, no. 6 (June): 613–629.

Steele, Claude M. and Joshua Aronson. 1995. "Stereotype Threat and the Intellectual Test Performance of African Americans." *Journal of Personality and Social Psychology* 69, no. 5: 797–811.

Steslow, K. 2010. "Metaphors in Our Mouths: The Silencing of the Psychiatric Patient." *The Hastings Center Report* 40, no. 4: 30–33.

Stoljar, Natalie. 2015. "'Living Constantly at Tiptoe Stance': Social Scripts, Psychological Freedom, and Autonomy." In *Personal Autonomy and Social Oppression: Philosophical Perspectives*, edited by Marina A. L. Oshana, 105–123. New York: Routledge.

Stoljar, Natalie. 2014. "Autonomy and Adaptive Preference Formation." In *Autonomy, Oppression, and Gender*, edited by Andrea Veltman and Mark Piper, 238–248. Oxford: Oxford University Press.

Stoljar, Natalie. 2011. "Informed Consent and Relational Conceptions of Autonomy." *Journal of Medicine and Philosophy* 36: 375–384.

Strawson, Peter. 2003. "Freedom and Resentment." In *Free Will*, 2nd ed., edited by Gary Watson, 72–93. Oxford: Oxford University Press.

Stuart, Heather. 2006. "Mental Illness and Employment Discrimination." *Current Opinion in Psychiatry* 19: 522–526.

Subica, Andrew M. and Bruce G. Link. 2024. "Mental Illness Stigma in Black, Latina/o, and Asian Americans." *Journal of Racial and Ethnic Health Disparities*: 1–12. https://doi.org/10.1007/s40615-024-02259-8.

Sue, Derald Wing, Christina M. Capodilupo, and Aisha M. B. Holder. 2008. "Racial Microaggressions in the Life Experience of Black Americans." *Professional Psychology: Research and Practice* 39, no. 3: 329–336.

Sue, Derald Wing, Christina M. Capodilupo, Gina C. Torino, et al. 2007. "Racial Microaggressions in Everyday Life." *American Psychologist* 62, no. 4: 271–286.

Sue, Derald Wing, Kevin L. Nadal, Christina M. Capodilupo, et al. 2008. "Racial Microaggressions against Black Americans: Implications for Counseling." *Journal of Counseling & Development* 86, no. 3: 330–338.

Sveinsdottir, Vigdis and Gary R. Bond. 2020. "Barriers to Employment for People with Severe Mental Illness and Criminal Justice Involvement." *Journal of Mental Health* 29, no. 6: 692–700.

Swistak, Zosia, Susan Sookoo, and Tom Jewell. 2024. "Integrating Subjective Recovery and Stigma Resistance in Individuals with Schizophrenia: A Narrative Review and Theoretical Integration." *Issues in Mental Health Nursing* 45, no. 5: 537–551.

Tartaro, Christine, Jess Bonnan-White, M. Alysia Mastrangelo, and Richard Mulvihill. 2021. "Police Officers' Attitudes toward Mental Health and Crisis Intervention: Understanding Preparedness to Respond to Community Members in Crisis." *Journal of Police and Criminal Psychology* 36: 579–591.

Tekin, Şerife. 2014. "Self-Insight in the Time of Mood Disorders: After the Diagnosis, Beyond the Treatment." *Philosophy Psychiatry, & Psychology* 21, no. 2 (June): 139–155.

Tekin, Şerife. 2011. "Self-Concept through the Diagnostic Looking Glass: Narratives and Mental Disorder." *Philosophical Psychology* 24, no. 3 (June): 357–380.

Thachuk, Angela K. 2011. "Stigma and the Politics of Biomedical Models of Mental Illness." *International Journal of Feminist Approaches to Bioethics* 4, no. 1 (Special Issue: Feminist Perspectives on Ethics in Psychiatry) (Spring): 140–163.

Thoams, Natalie and Helena Menih. 2022. "Negotiating Multiple Stigmas: Substance Use in the Lives of Women Experiencing Homelessness." *International Journal of Mental Health and Addiction* 20: 2973–2992.

Thornicroft, Graham. 2009. "Global Pattern of Experienced and Anticipated Discrimination against People with Schizophrenia: A Cross-Sectional Survey." *The Lancet* 373 (January): 408–415.

Thornicroft, Graham. 2006. *Shunned: Discrimination against People with Mental Illness*. Oxford: Oxford University Press.

Thornicroft, Graham, Charlene Sunkel, Akmal Alikhon Aliev, et al. 2022. "The *Lancet* Commission on Ending Stigma and Discrimination in Mental Health." *Lancet* 400, no. 10361: 1438–1480.

Thornton, Tim. 2020. "The Implications of the Loss of Self-Respect for the Recovery Model in Mental Healthcare." *Human Affairs* 30, no. 3: 316–327.

Tiderington, Emmy, Benjamin F. Henwood, Deborah K. Padgett, and Bikki Tran Smith. 2020. "Employment Experiences of Formerly Homeless Adults with Serious Mental Illness in Housing First versus Treatment First Supportive Housing Programs." *Psychiatric Rehabilitation Journal* 43, no. 3: 253–260.

Tlachac, M. I., Miranda Reisch, Brittany Lewis, et al. 2022. "Impact Assessment of Stereotype Threat on Mobile Depression Screening Using Bayesian Estimation." *Healthcare Analytics* 2: 1–14. Accessed January 11, 2025. https://doi.org/10.1016/j.health.2022.100088.

Torales, Julio, Telmo Raúl Aveiro-Róbalo, Carlos Ríos-González, et al. 2023. "Discrimination, Stigma and Mental Health: What's Next?" *International Review of Psychiatry* 35, nos. 3–4: 242–250.

United Nations. 1966. "International Covenant on Economic, Social and Cultural Rights." *United Nations Human Rights Office of the High Commissioner* (adopted 16 December). Accessed February 10, 2024. www.ohchr.org/en/instruments-mechanisms/instruments/international-covenant-economic-social-and-cultural-rights.

U.S. Centers for Disease Control and Prevention. 2024. "Symptoms of Anxiety and Depression among Adults: United States, 2019 and 2022." *National Health Statistics Reports* 213 (November 7): 1–14. Accessed December 26, 2024. www.cdc.gov/nchs/data/nhsr/nhsr213.pdf.

U.S. Department of Housing and Urban Development (HUD). 2017. "HUD Report Finds People Living with Mental Disabilities Face Significant Rental Housing Discrimination." *HUD Archives: News Releases* (September 5). Accessed August 22, 2024. https://archives.hud.gov/news/2017/pr17-070.cfm.

Valery, Kevin-Marc, Louis Violeau, Thomas Fournier, et al. 2023. "Part of the Solution Yet Part of the Problem: Factors of Schizophrenia Stigma in Mental Health Professionals." *International Journal of Psychiatry in Clinical Practice* 27, no. 2: 134–144.

Veroniki Karidi, Maria, Costas N. Stefanis, Christos Theleritis, et al. 2010. "Perceived Social Stigma, Self-Concept, and Self-Stigmatization of Patient with Schizophrenia." *Comprehensive Psychiatry* 51: 19–30.

Villatoro, Alice P., Melissa J. DuPont-Reyes, Jo C. Phelan, and Bruce G. Link. 2022. "'Me' versus 'Them': How Mental Illness Stigma Influences Adolescent Help-Seeking Behaviors for Oneself and Recommendations for Peers." *Stigma and Health* 7, no. 3: 300–310.

Voldby, Katrine Gren, Lone Christina Hellström, Marie Eva Berg, and Lene Falgaard Eplov. 2022. "Structural Discrimination against People with Mental Illness; A Scoping Review." *SSM-Mental Health* 2: 1–20 (Article 100117).

Waghorn, Geoff and Chris Lloyd. 2005. "The Employment of People with Mental Illness." *Australia e-Journal for the Advancement of Mental Health* 4, no. 2 (Supplement): 1–43.

Wahl, Otto F. 2012. "Stigma as a Barrier to Recovery from Mental Illness." *Trends in Cognitive Sciences* 16, no. 1 (January): 9–10.

Walton, Gregory M. and Geoffrey L. Cohen. 2007. "A Question of Belonging: Race, Social Fit, and Achievement." *Journal of Personality and Social Psychology* 92, no. 1: 82–96.

Wan, William. 2022. "Yale Accused of Discriminating against Students with Mental Illness in Lawsuit." *The Washington Post* (November 30).

Wang, Zixin, Charles Chiu Hung Yip, Donald Chi Kin Leung, and Kevin Ka Shing Chan. 2023. "The Impact of Mindfulness on Stigma Stress and Well-Being among Individuals with Mental Disorders." *Mindfulness* 14: 808–817.

Waqas, Ahmed, Salma Malik, Ania Fida, et al. 2020. "Interventions to Reduce Stigma Related to Mental Illnesses in Educational Institutes: A Systematic Review." *Psychiatric Quarterly* 91, no. 3: 887–903.

Wardrope, Alistair. 2015. "Medicalization and Epistemic Injustice." *Medicine, Health Care, and Philosophy* 18: 341–352.

Watson, Amy C. and Beth Angell. 2013. "The Role of Stigma and Uncertainty in Moderating the Effect of Procedural Justice on Cooperation and Resistance in Police Encounters with Persons with Mental Illnesses." *Psychology, Public Policy, and Law* 19, no. 1: 30–39.

Watson, Amy C., Beth Angell, Theresa Vidalon, and Kristin Davis. 2010. "Measuring Perceived Procedural Justice and Coercion among Persons with Mental Illness in Police Encounters: The Police Contact Experience Scale." *Journal of Community Psychology* 38, no. 2: 206–226.

Watson, Gary. 2003. "Free Agency." In *Free Will*, 2nd ed., edited by Gary Watson, 337–351. Oxford: Oxford University Press.

Watters, Ethan. 2010. *Crazy Like Us: The Globalization of the American Psyche.* New York: Free Press.

Weisz, Carolyn and Diane M. Quinn. 2018. "Stigmatized Identities, Psychological Distress, and Physical Health: Intersections of Homelessness and Race." *Stigma and Health* 3, no. 3: 229–240.

West, Keon, Miles Hewstone, and Simon Lolliot. 2014. "Intergroup Contact and Prejudice against People with Schizophrenia." *The Journal of Social Psychology* 154: 217–232.

Westlund, Andrea C. 2009. "Rethinking Relational Autonomy." *Hypatia* 24, no. 4: 26–49.

Westlund, Andrea. 2018. "Relational Autonomy and Practical Authority." In *The Bloomsbury Companion to Analytic Feminism*, edited by Pieranna Garavaso, 375–393. London: Bloomsbury.

Whisler, Adam, Naheed Dosani, Matthew J. To et al. 2021. "The Effect of a Housing First Intervention on Primary Care Retention among Homeless Individuals with Mental Illness." *PLOS One* 16, no. 2 *(February)*: 1–12. https://doi.org/10.1371/journal.pone.0246859.

Wilton, Robert and Joshua Evans. 2018. "Accounting for Context: Social Enterprises and Meaningful Employment for People with Mental Illness." *Work* 61: 561–574.

Wong, Eunice C., Rebecca L. Collins, Jennifer L. Cerully, Jennifer W. Yu, and Rachana Seelam. 2018. "Effects of Contact-Based Mental Illness Stigma Reduction Programs: Age, Gender, and Asian, Latino, and White American Differences." *Social Psychiatry and Psychiatric Epidemiology* 53, no. 3: 299–308.

Young, Iris Marion. 2011. *Responsibility for Justice*. Oxford: Oxford University Press.

Yu, Ben C. L., Winnie W. S. Mak, and K. H. Lai. 2025. "Testing the Interconnectedness Hypothesis: A Longitudinal Examination of the Association between Interconnected Explanation and Accountability on Mental Illness Stigma." *Stigma and Health* 10, no. 3: 1–7 (advance online publication). https://dx.doi.org/10.1037/sah0000499.

Zautra, Alex J., John Stuart Hall, and Kate E. Murray. 2010. "Resilience: A New Definition of Health for People and Communities." In *Handbook of Adult Resilience*, edited by John W. Reich, Alex J. Zautra, and John Stuart Hall, 3–29. New York: Guilford Press.

Zimmerman, David. 2003. "Sour Grapes, Self-Abnegation and Character Building: Non-Responsibility and Responsibility for Self-Induced Preferences." *The Monist* 86, no. 2: 220–241.

Index

Printed in the United Kingdom by TJ Clays Ltd.